T0074602

Climate Change Ethics for an Endangered World

Climate change confronts us with our most pressing challenges today. The global consensus is clear that human activity is mostly to blame for its harmful effects, but there is disagreement about what should be done. While no shortage of proposals from ecological footprints and the polluter pays principle to adaptation technology and economic reforms, each offers a solution – but is climate change a problem we can solve?

In this provocative new book, these popular proposals for ending or overcoming the threat of climate change are shown to offer no easy escape and each rest on an important mistake. Thom Brooks argues that a future environmental catastrophe is an event we can only delay or endure, but not avoid. This raises new ethical questions about how we should think about climate change. How should we reconceive sustainability without a status quo? Why is action *more* urgent and necessary than previously thought? What can we do to motivate and inspire hope? Many have misunderstood the kind of problem that climate change presents – as well as the daunting challenges we must face and overcome. *Climate Change Ethics for an Endangered World* is a critical guide on how we can better understand the fragile world around us before it is too late.

This innovative book will be of great interest to students and scholars of climate change, climate justice, environmental policy and environmental ethics.

Thom Brooks is Professor of Law and Government, the Dean of Durham Law School and Associate Member of the Philosophy Department and School of Government and International Affairs at Durham University. He is founding editor of the *Journal of Moral Philosophy*.

Routledge Focus on Environment and Sustainability

Jainism and Environmental Politics
Aidan Rankin

Australian Climate Policy and Diplomacy
Government-Industry Discourses
Ben L. Parr

Reframing Energy Access
Insights from The Gambia
Anne Schiffer

Climate and Energy Politics in Poland
Debating Carbon Dioxide and Shale Gas
Aleksandra Lis

Sustainable Community Movement Organizations
Solidarity Economies and Rhizomatic Practices
Edited by Francesca Forno and Richard R. Weiner

Climate Change Ethics for an Endangered World
Thom Brooks

The Emerging Global Consensus on Climate Change and Human Mobility
Mostafa M. Naser

For more information about this series, please visit: www.routledge.com/Routledge-Focus-on-Environment-and-Sustainability/book-series/RFES

Climate Change Ethics for an Endangered World

Thom Brooks

First published 2021
by Routledge
2 Park Square, Milton Park, Abingdon, Oxon OX14 4RN

and by Routledge
52 Vanderbilt Avenue, New York, NY 10017

Routledge is an imprint of the Taylor & Francis Group, an informa business

British Library Cataloguing-in-Publication Data
A catalogue record for this book is available from the British Library

Library of Congress Cataloging-in-Publication Data
A catalog record for this book has been requested

ISBN: 978-0-367-52431-9 (hbk)
ISBN: 978-1-003-05795-6 (ebk)

Typeset in Times New Roman
by Apex CoVantage, LLC

Printed in the United Kingdom
by Henry Ling Limited

For my daughter, Eve Brooks

Contents

Preface

Each book tells its own story. This work is no different. I have been interested and deeply concerned about climate change and its effects since starting my studies in philosophy as an undergraduate student, and possibly earlier. My childhood home was built in the mid-1860s, but now found itself in a flood zone that gave us something extra to worry about whenever major storms came to the New Haven, Connecticut area. Visualizing rising waters around our home during occasional hurricanes brings back vivid memories. In New London, Connecticut, a short drive away, hurricane protection barriers were constructed guarding against increasing coastal erosion during that time as well. Climate change and its effects have been with me for most of my life. It is nothing new – and I have never lived in a flood zone again.

As I draft this Preface and complete this manuscript, it is early May 2020 and both the UK and USA are in the midst of fighting a global pandemic whose roots have been associated with climate change, too. The coronavirus COVID-19 exercises a profound effect on how we relate to each other and to nature. While the current outbreak is still a matter of study, this book will offer some insights into its harmful impacts and the new opportunities it creates for the future. Every indication thus far while living in quarantined lockdown is that life and our livelihoods will not be the same again.

While my interest and concern in this area have always been high, I did not begin writing about climate change issues until much later. The main catalyst is an unexpected consequence of hosting a seminar series over several years. When I started my first position at Newcastle University, they had unwisely closed the philosophy department years before my arrival. So I set myself a first task to establish a new seminar series in political and legal philosophy to recreate some of the research environment that had gone lost. It was relatively easy to do building on the terrific foundations of some brilliant colleagues across several departments like Peter Jones, Richard Mullender and the late Mary Midgley who, although she was retired, Mary was incredibly active and all eager to see a vibrant philosophical culture thrive once more.

The fortnightly Newcastle Ethics, Legal and Political Philosophy seminar series I hosted had no theme, but a frequent topic of many papers was climate change ethics. I invited some of the very best in the field who gave insightful talks on how we should think about the problem and often proposed solutions. Big problems can be *big* for a reason; namely, the challenge is immense and does not admit of any one solution if only we did one thing or another. Climate change seemed to me to be such a problem – and claims that primarily the use of ecological footprints, polluter pays principle or placing our faith in future technologies could make our worries go away seemed to all not hit their mark. This is despite the fact I was desperately interested to become convinced by one of the many options on the table. It was not to be.

Over the years, I found myself regularly making the same kinds of objections to papers by leading figures, most of which was later published and examined in this book. However much I admired their work, and I deeply admire it or otherwise I would not have spent the time in writing a book about it, I was often left unsatisfied by the replies.

So you might say that this book is either the product of much of my lifelong interest in and concerns about climate change – or my finally spelling out at book length my concerns about many papers delivered in forums like my former political and legal philosophy group over several years, and then reworked for many more years afterwards.

Since 2012, I moved institutions and departments. After I earned a Ph.D. in Philosophy while working in Politics, I moved to a law school – and a brilliant one at that – at Durham University nearly eight years ago. For half this time, I have been deeply honored to become Durham Law School's inaugural Dean. My colleagues in Law and departments where I have associate membership – Philosophy and the School of Government and International Affairs – are a source of much inspiration for me and I am enormously grateful to them for their firm support for my work on this and other matters.

Originally, this book came to life as a target article: Thom Brooks, "How Not to Save the Planet," *Ethics, Policy and Environment* 19(2) (2016): 119–135. I am most grateful to the publisher for permission to include it in this book. I am deeply appreciative to the editors of *Ethics, Policy and Environment*, especially Benjamin Hale, and their referees for helpful comments and putting my work in the spotlight. I am further grateful to Andrew Jameton, Jordan Kincaid, Alexander Lee, Clement Loo, Ben Mylius, Eoin O'Neill and Jordan Peter Schwartz for their terrific commentaries on my original article. I offer responses to their penetrating and constructive criticisms in the text below. I must also strongly highlight that this book significantly revises, updates and expands my original account.

Versions of the original piece and the current manuscript have been presented at a long list of department seminars, including the universities of Boston, Cardiff, Durham, Edinburgh, Essex, Groningen, Oxford, Oxford Brookes, Stirling and Yale. This work has also been presented at the annual Oxford Political Thought Conference and the annual Political Studies Association meeting. Additionally, I had the pleasure of presenting this work to the Edinburgh Fringe Festival. I am most grateful to these audiences and others for helpful comments on earlier drafts, most especially Robin Attfield, Andrea Baumeister, Richard Bellamy, David Boucher, Gillian Brock, Claire Brooks, Gary Browning, Carolyn Cole Candolera, Simon Caney, Alan Carter, Hugh Compston, James Connolly, Rowan Cruft, Boudewijn de Bruin, Liz Fraser, Fabian Freyenhagen, Mathias Frisch, Les Green, Bruce Haddock, Iain Hampsher-Monk, Nicole Hassoun, Clare Heyward, John Gardner, Peter Hulm, Peter Jones, Pauline Kleingeld, Joshua Knobe, Melissa Lane, S. Matthew Liao, Graham Long, Jonathan (E.J.) Lowe, Sandra Marshall, Wayne Martin, Liz McKinnell, David Miller, Margaret Moore, Richard Mullender, Aletta Norval, Martha Nussbaum, Jörgen Ödalen, David Owen, Soran Reader, Peri Roberts, David Schlosberg, Thomas Schramme, Esther Shubert, Peter Singer, Matthew Noah Smith (to whom I owe thanks for convincing me my original article should not be called "How to Save the Planet" but "How *Not* to Save the Planet" at a meeting of the Yale Moral Philosophy Workshop), Suzanne Sreedhar, Daniel Star, Martin van Hees, Margaretha Wewerinke, Jo Wolff, Hiro Yamazaki and Lea Ypi. My apologies to any that I have left out.

Part of the writing of this book was done during my time as a visitor to Columbia University's Law School, New York University's Center for Bioethics and the University of Chicago Law School. I am most grateful to colleagues for the warm and friendly environment to work on the manuscript, including Adam Kolker and Matthew Liao who made these opportunities possible.

I owe a very special – and large – debt to Chicago's Martha Nussbaum, who has always been a huge source of inspiration and support for me since my graduate student days, as well as a dear friend. Her work, example and terrific advice both personal and professional have always guided my thinking and informs much of the critical perspective advanced in this book.

My thanks must also go to Routledge's Alison Kirk for encouraging me to write this book and to Annabelle Harris for guiding this manuscript through to publication. It has been a delight to work with their team.

I am enormously grateful to my family, especially to Claire Brooks, for their support which too often requires my being locked away for many hours studying papers and typing away instead of other tasks.

I dedicate this book to my young daughter, Eve Brooks. When I reflect on the future of our planet, I think about her future and that of her generation.

Climate change is happening now and the challenge of addressing it is greater than often claimed. Our commitment to a sustainable future in an endangered world must rise to this great challenge. I hope this book offers some light – for Eve and her generation – at a time of much darkness. Light must always prevail and, if we can retain hope for a sustainable future, it will.

T.A.K.B.

Durham, England

Introduction

"Global warming, climate change, the devastating loss of biodiversity are the greatest threats that humanity has ever faced and one largely of our own creation."

– Prince Charles

No one is an island

Hilia's world is changing fast. Her island home is reliant on collecting rainwater or importing fresh water to drink.[1] The land has become so salty from rising sea levels that foaming salt can froth as it pokes through the porous soil now all but useless for planting crops.[2] What does grow, like bananas, can fail to ripen in the harsh conditions.

Her home is slowly, but surely, sinking under the waves. This has led to an increasing dependency on importing food. She says: "What I tell our people is you have to rely on what you see. You see the pulaka pits getting salty and the banana palms dying. You see the storms."[3] In short, what you see are the effects of climate change on our endangered world.

Hilia lives in Tuvalu, the fourth smallest country in the world. A thin, sandy sliver in the Pacific Ocean, the country is located about hallway between Hawaii and Australia. It has been inhabited for over 2,000 years. Most of Tuvalu's islands rise no more than about ten feet above sea level. Once a launch pad for the United States during World War II, these islands are beginning to disappear off the map.

Other Pacific islands are not so lucky. Several of the Solomon Islands and parts of Micronesia have submerged completely. All that remain are oral stories about once proud kingdoms now vanished.[4]

These areas are under an existential threat impacting on their people, wildlife and local ecosystems. For local residents like Hilia, climate change is neither controversial nor an abstract problem. It is part of a lived experience –

and it is on course to drown her home over the next few decades. Doing nothing means many parts of our planet's land mass will have nothing left of their people, their ways of life and the natural habitat. A world will be lost.

Tragically, this bleak outlook is their future through no fault of their own. While these peoples may live on islands, they are not separated from global climatic forces linking us all together. Greenhouse gas emissions and the environmental damage they create can outlive a generation – and their harmful effects can last for a generation or more for their areas and impact on those living all across our planet. You cannot isolate yourself from the effects of climate change wherever you may life. No one is an island; all of us are interconnected. We share a single home: the Earth.[5]

Hilia's drowning Tuvalu is only one of many vivid examples. And it is *drowning*. This island country is becoming swallowed up by the ocean waves more each year. It is not because the islands are crumbling into the sea. Tuvalu is disappearing because the ocean water around it is rising – and this is because of human activities affecting the climate. Tuvalu is not a sinking nation, but a country being drowned.

Humanity displaced

Conflict and violence are no longer the main causes of people leaving their homes and homelands. In 2018, the International Organization for Migration found that over 60 percent of all forced migration – or about 17 million people – occurred because of natural disasters.[6] People being displaced because of adverse and extreme weather is the new normal – and a growing humanitarian crisis.

Osman fled her home in Somalia for safety in neighboring Kenya last year. A severe drought had led to a lack of drinking water, crops had withered and all local livestock had died. Leaving her home was not a choice, but a necessity if she and her baby were to survive. She said: "It's not that we are running away from conflict and war. We are running away from hunger."[7] Relocation was a form of self-defense against a very real environmental threat.

Droughts used to come every seven or more years. Now they arrive almost every other year. A fellow Somali refugee to Kenya said that you know things are bad when even the camels die from thirst.[8] Over two million Somalis were at risk of starvation – and a changing climate transforming an average rainy summer season into one of the driest on record for more than 35 years.[9] In Somalia, the droughts are so bad they are named. The last one was called *Sima*, or "equal," because it hit everyone equally hard.[10]

Fast forward to our future. If the climate heats up by an extra 2°C by 2100, the only way to visit Miami will be by boat with six million living in Florida at serious risk from a rising sea level.[11] The traditional song

America the Beautiful speaks of a land of "spacious skies," a "fruited plain" and "brotherhood from sea to shining sea."[12] All are at risk this century with major cities coast to coast from New York and New Orleans to even parts of San Francisco gradually submerging under a rising tide.[13] Everything will be affected everywhere we look.

Our greatest challenge today

Climate change confronts us with our most urgent and greatest challenges today.[14] It threatens nonhuman species with extinction – and our survival as well. The effects are with us now, from Hilia's Tuvalu or Osman's Somalia today to many other countries in the future. Climate change is happening, the developed world will be impacted as will developing countries and the challenges they pose for us cannot be avoided or ignored.

Our planet is changing – and it is important to understand how and why. Consider these facts. Roughly ten percent of the Earth's land is covered by ice sheets or glaciers.[15] These are melting because the surface air temperature in areas like the Arctic has more than doubled the global average for two decades running – and the Antarctic's ice sheet has seen accelerated losses over the last decade, too.[16]

When this ice melts, it damages unique habitats threatening some animal and plant species with extinction – while also raising global sea levels.[17] The sea has risen by about six inches since 1902. By 2100, the sea level is expected to rise between 1.3 to 2.6 feet.[18] As our oceans have warmed, they have seen a general decrease in oxygen to a depth of 1,000 meters.[19] This has changed the abundance and composition of species as the marine ecosystem is impacted by these changes.[20] The rising tide threatens places like Tuvalu, but also coastal communities around the world as they become more vulnerable to flooding.

We are witnessing these rising sea levels with an increasing frequency of extreme weather events.[21] This is caused by the increasing volume of water unlocked from melting ice sheets and glaciers. It contributes to heavier downfalls, more dangerous hurricanes and a shifting jet stream with changing weather patterns. Not all areas are heated or affected the same, with tropical areas warming more quickly than elsewhere.[22] These factors have had a significant impact, including the greater likelihood of droughts affecting agricultural production, an increase in wildfires, the spread of tropical diseases, such as water-borne illnesses, to new geographical regions and the more recent phenomena of "environmental refugees" as more people leave their drowning homelands.[23]

Climate change is about more than hotter weather. So many nonbiodegradable plastics have been dumped in our oceans and waterways that there is now a 'Great Pacific Garbage Patch' of microplastics.[24] This stretches from the West Coast of America to Japan. The United Nations Environmental Program estimates that for every square mile of ocean there are about 46,000 pieces of plastic.

As this book is being written, over eighteen million people have contracted the coronavirus COVID-19, including more than 700,000 people who have died in a global pandemic.[25] The outbreak has led to lockdowns across the world lasting months and stretching national health care systems to their breaking points. While many countries implemented record levels of spending packages, millions have lost their jobs and livelihoods while tens of thousands have lost their lives.

The source of the virus is still unclear although scientists have identified bats as a possible cause.[26] However, bats are nocturnal and reclusive, raising questions about how this disease was transferred from one species to ours. The rapid spread of human beings into new areas impacting on, if not destroying, natural habitats is one possible explanation for the origins of the current global pandemic – and its deadly consequences worldwide. Our climate is changing in our land use as well as through fossil fuel emissions.

When I was a university student in the 1990s, people commonly spoke of "global warming" but it is more accurate to call it "climate change," as it is most often described today. Global warming specifically refers to the Earth's rising average temperature since about 1850 because of human activities like burning fossil fuels, such as coal, oil or natural gas.[27] This was first called *the greenhouse effect* in 1896 by Svante Arrhenius, a Swedish chemist. Greenhouse gasses like carbon dioxide, methane, nitrous oxide and industrial gases like chlorofluorocarbons trap additional infrared radiant energy within our atmosphere and this warms it.[28] As David Weisbach explains: "as the greenhouse gas blanket gets thicker, we get warmer."[29] A hotter planet is certainly one product of climate change that comes readily to mind when we see effects like melting ice caps leading to rising sea levels and threatened coastlines, for example.

Climate change is much more than global warming, as the current global pandemic puts into focus. As more water is unlocked from ice caps, more moisture becomes mobilized within the atmosphere. This extra moisture fuels the more frequent and intense extreme weather, including ice storms as there is more water in the atmosphere that can become snow – with a shifting jet stream to boost it.[30] So a warming planet overall does not mean that less snow will fall. (And this explains why U.S. President Donald J. Trump is wrong to think more severe winter storms are evidence against climate change. In fact, they help confirm this change.[31])

Climate change is about our altering and endangered planet in the broadest sense. This includes the melting ice and rising seas, but varying and more extreme weather patterns, increase in droughts and wildfires and the spread of diseases to new areas. But climate change also refers to deforestation, our land uses including agricultural-related activities, geoengineering and genetically modified crops, sanitation and more.[32] These changes have a profound effect on wildlife, but also the way in which we interact with the natural environment. Human beings impact climate change in more ways than the atmosphere alone, but also on land and by sea in how we interact with inhabitable spaces for ourselves, our uses of transport and the waste we leave behind.

There is no doubting the science behind climate change. In 1988, the United Nation's Environment Program and the World Meteorological Organization joined forces to establish an authentic voice of the global scientific consensus: the United Nation's Intergovernmental Panel on Climate Change (IPCC).[33] The scientific community is clear. IPCC climate change reports confirm that "warming of the climate system is unequivocal" and not in any doubt.

These reports further establish a clear, global scientific consensus that human beings are primarily responsible for climate change and its associated dangers.[34] Climate change is a phenomenon forged by human hands. This is caused by factors like the creation of greenhouse gasses, such as carbon emissions. For example, it has been at least 420,000 years since the Earth has had so much carbon dioxide and methane in its atmosphere.[35] If all emissions were reduced instantly to zero today, there is already enough carbon emissions to continue global warming for decades.[36] It is estimate that 15 percent to 40 percent of CO_2 emissions until 2100 will remain in the atmosphere for over 1,000 years – and 10 percent to 25 percent of CO_2 emissions will linger for 10,000 years.[37]

The damage we do to our environment outlives us, impacting on future generations. None of this is in doubt. Some now claim that the current geological epoch, the Holocene, covering over 10,000 years should be renamed *the Anthropocene*. This would recognize, if only in name, the profound geological impact human beings have on the planet as our impact becomes "the equal of tectonic shifts, glaciation, volcanic eruption, asteroid collisions and the like."[38]

The only disagreement among the global scientific community is about various predictions for exactly how much sea levels will rise or the precise rise in temperature over time – there is no disagreement about the fact that these *will happen* and primarily because of *human activities*.[39] There are variances in predictions about the future based on different models of expected emissions and their impact. However, the broad consensus about the fact of climate change and its effects thus far is remarkably far stronger than any expert agreement in other fields like finance or even health.[40] The tide is rising and the environment is changing, but how fast and by how much will depend on future behaviors as well as accounting for our past and present. Predicting the future with exact precision is difficult because it is aiming at a moving target. But that does not mean we are powerless.

From the IPCC's reports establishing the scientific consensus supporting climate change, the global community came together to agree to the United Nations Framework Convention on Climate Change (UNFCCC) in 1992 with nearly 200 signatories, including China, the United Kingdom and the United States.[41] This required "stabilization of greenhouse gas concentrations in the atmosphere at a level that would prevent dangerous anthropogenic interference" and agreed on "common but differentiated responsibilities" among states whereby the most affluent would lead in cutting emissions.[42]

Yet despite the fact there is widespread acceptance of the fact and causes of climate change – which the broad support for the UNFCCC and IPCC climate change reports attest to – there is no consensus about what to do about the clear and present danger we find ourselves in. No wonder some claim we are living amidst a climate emergency requiring urgent and substantive action.[43] Many believe the dangers related to climate change pose the greatest problems for governments today.[44] I believe these campaigners are right.

This chapter starts with a brief epigraph from Prince Charles. It is intentional. For over four decades, he has been both a tireless champion for sustainability and one of the first major world leaders to accept the science advocating for urgent change.[45] What was pioneering – even considered by some as extreme – in the past has become everyday common sense today. This transition did not happen overnight. Campaigns can make a difference in raising awareness, as the issue of climate change shows clearly.

So our challenge is clear: it is not to consider *whether there is* climate change, but instead *what to do* and best respond to it.[46] Human-caused climate change confronts us with such grave threats, but is this a problem we can solve? This is the issue we will focus on in this book.[47]

Two possible solutions in search of a problem: mitigation and adaptation

The answer for many academics and policy-makers is a clear *yes* – climate change is a problem that we can solve. Examples abound from Nicholas Stern's *A Blueprint for a Safer Planet: How We Can Save the World and Create Prosperity* to former U.S. Presidential candidate and New York Mayor Michael Bloomberg's coauthored *Climate of Hope: How Cities, Businesses and Citizens Can Save the Planet*.[48] Such views see climate change as not beyond our control; our domination of nature has been destructive thus far or it can also be reconstructive by creating a sustainable future.

The only serious disagreement to be found is about *how* we should do this. These solutions to the challenges of climate change can be broadly divided into two overlapping camps that are not mutually exclusive of the other. Both aim to effectively overcome climate change-related problems to ensure that they do not lead to our planet becoming inhospitable for us.[49]

The first camp prioritizes *mitigation*. Climate change is happening and so the goal is broadly conservationist in mitigating the harmful effects. This goal is to be achieved by reducing carbon emissions and, thus, better manage climate change's effects by decreasing, if not ending, contributions to any further climate changes.[50]

Mitigation supporters advocate policies like using an *ecological footprint*, a measure of environmental impact, as a means for setting clear limits to permissible carbon emissions. This policy relates to *carbon trading* where states trade shares of permissible emissions within a global cap binding on all. Finally, there is the popular policy of the *polluter pays principle* whereby those who create emissions which contribute to climate change are required to pay more with a view to these extra costs reducing overall emissions and using the additional funding to implement mitigation measures to address environmental impacts new and old.

The second camp prioritizes *adaptation*. Climate change is happening and so their goal is to better adapt ourselves to our changing environment. We can become more effectively protected and overcome the potentially damaging consequences of climate change in that way. Policies in this area emphasize technological advancements, such as flood defenses to protect against coastal flooding and geoengineering genetically-modified crops to better withstand droughts and worsening conditions for growing food.

While most advocates for policies that either primarily promote mitigation or adaptation, most proposals incorporate some combination of both.[51] For example, mitigation proponents often acknowledge the need for some adaptive measures. While we can make a significant difference on how we contribute (or not) to further climate changes, some such changes are happening already which need to be addressed here and now. Likewise, most adaptation proponents champion the strides forward science and technology have made over a relatively short period of time. Nonetheless, this progress has limits and reducing our impact through cutting greenhouse gas emissions makes this difficult task somewhat easier. Otherwise, technological advances might not have a chance at succeeding.

So while there are two different ways of prioritizing different goals – whether it is mitigation or adaptation – they are not mutually exclusive. For most proponents of these views, some combination of mitigation and adaptation can help us "solve," in their eyes, the problem of climate change and its harmful effects. This has contributed to various policy proposals that have been championed as potential silver bullets to bring climate change's problems to an end. Their goal of creating the conditions for a more happy ever-after – if only we lived within an equal ecological footprint, make polluters pay if they pollute or create the right adaptive technologies – have understandable widespread appeal among politicians, policy-makers, the public and even many of my fellow university professors. A happy ending is tempting, even desirable, but nonetheless appears to be beyond our grasp.

These creative ways of reducing our environmental impact and improving our ability to thrive amidst changing climatic conditions are important contributions to how we should think about the future. Yet, they can suffer from making a false assumption about the problem they each claim to "solve."

Climate change is an unavoidable challenge

In this book, I challenge the orthodox consensus that climate change is a problem that can be solved once and for all. A future ice age or other environmental catastrophe is an event that we might at best *delay*, but *not avoid* forever.

Our planet has suffered natural catastrophes before human beings evolved. That is a point too often overlooked. While human activities may contribute to making such a disastrous calamity happen sooner rather than later, this raises new questions for climate change ethics for our endangered world. We must ask: what are the moral implications of a future climatic catastrophe that might be delayed at best? What practical consequences might these implications yield? These concerns are too often neglected in favor of defending solutions that fail to end the serious effects of climate change which they promise.

There is "a perfect moral storm" whirling around us, as described powerfully by philosopher Stephen Gardiner.[52] Affluent countries may be tempted to pass on the costs of climate change to poorer states. Present generations may be tempted to pass on the problem to future generations to solve. Global inaction is fueled by a poor grasp of how science and international justice connect with our relationship to the natural environment. With so much at stake, there are many reasons for little to be done now or in sufficient time. This in no way supports our doing nothing, but only highlights the significant obstacles to our making progress.

An unnoticed dilemma in this perfect moral storm is how the problem to be "solved" is understood. At the crux of this is what I call *the problem of end-state solutions*.[53] This is a problem, or perhaps we might liken it instead to a myth, of believing that *if only* carbon emissions were below a forever fixed threshold – or *if only* we had a particular technological advance adapting us to a current climatic conditions – that any further climate change in future might be prevented and so the problems associated with climate change would be ended. In other words, a solution might be thought to fix the problem because the latter is not a moving target, but rather it is set in stone-like end-state.

I will argue that this problem, or myth, is overlooked at our peril. Climate change did not require human beings for it to happen before, even if human activity is the cause of our biggest challenges today. Future climate changes will not be prevented whatever our actions. My view accepts the full scientific consensus, but raises serious doubts about proposed "solutions."

However, it would be a mistake to claim that we should do nothing about climate change and become resigned to a future where environmental catastrophe is on our horizon awaiting us a like a ticking time bomb. Swift action now can postpone such an event further into the future and buy time for mitigation and adaptation strategies to help us face this future as best as possible. My claim that proposed "solutions" mistake the nature of their problem is not a call for pessimism or complacency. I am a realist and an optimist. We live within an endangered world and this calls for a different set of ethical considerations for thinking about how best to respond to a challenge that will never fade from view entirely. When inspiring young activists like Greta Thunberg warn us that we are confronting a climate emergency, they are right – and the implications of this are more significant – and sobering – than might be readily apparent for current and future generations alike.

Plan for this book

Let me briefly summarize the following chapters for this book.

The next two chapters critically examine mitigation strategies. The first considers the defense of the ecological footprint and its offshoots. This position claims broadly that there is a sustainable ever-after of carbon emissions the planet can absorb. The amount of permissible carbon for our planet is divided equally by the world's population to be managed by their governments and international organizations. The second chapter examines a different approach to mitigation called the polluter pays principle. It claims that polluters should pay extra to disincentivize continued high emissions by making this more expensive and using the new revenue to address damage caused. Each of these mitigation strategies will be presented to show why they are attractive options for their supporters. But it will be argued that neither is a silver bullet that can "end" the problem they seek to fix.

The third chapter turns to adaptation. Its adherents claim that advances in science and technology can sufficiently protect human beings from a changing climate and its potentially dangerous effects. Much of the arguments for this view rest both on untested or not yet invented technological innovations alongside a belief that there is possible to forever more technologically adapt to our changing environment. It will be argued that this belief is built on a hope that such advances will – and, crucially, must – be achieved without room for any error where a mistake could signal a hastening catastrophe.

The fourth chapter turns to climate change and catastrophe more broadly. The previous chapters aim to show how policy proposals have importance, but none offer a solution. This fourth chapter explains why the inevitability of climate change is a fact, but not a reason for treating our current circumstances as anything short of a climate emergency requiring our addressing new ethical questions and demanding more radical action.

The fifth chapter considers a range of possible objections. Some of these respond to criticisms of my previously published work.[54] But I also consider other potential criticisms as well. The book then ends with a conclusion tying the different strands together in a short summary. I argue for a new understanding of the kind of challenge that climate change presents. A clearer understanding of this challenge highlights the need to conceive of climate change as a problem to be managed and perhaps never solved. I conclude considering the possible implications for our theories about climate change justice and how this might lead to action.

Climate change is happening. Human beings are its primary cause. Doing nothing is not an option. But *what* should we do – and *why* should we do it? This book will now turn to trying to unravel and answer these important questions.

Notes

1 See Eleanor Ainge Roy, "'One Day We'll Disappear': Tuvalu's Sinking Islands," *The Guardian* (May 16, 2019), url: www.theguardian.com/global-development/2019/may/16/one-day-disappear-tuvalu-sinking-islands-rising-seas-climate-change. Where free online resources are available, these will be included in footnotes. All links "live" as of the completion of this manuscript on April 30, 2020.
2 See Roy, "One Day We'll Disappear" and Kerri MacDonald, "As Water Rises, There's No Place Like (or for) Home," *New York Times* (October 18, 2011), url: https://lens.blogs.nytimes.com/2011/10/18/no-place/.
3 See Kennedy Warne, "That Sinking Feeling," *New Zealand Geographic* 70 (November–December 2004), url: www.nzgeo.com/stories/tuvalu/.
4 See Patrick D. Nunn, "Sinking Islands: Sea Level Rise Is Washing away Micronesia's History," (September 11, 2017), url: www.newsweek.com/sea-level-rise-vanishing-islands-micronesia-history-706455.
5 In this book, I will refer to "us," "we," "our" and so on. This naturally raises the question of who "we" are. While this book is meant to speak across cultures and continents, the use of "we" will often be directed towards those of us living in affluent, democratic Western societies which have a great responsibility for acting urgently to address the current climate crisis. "We" have much to do and fast.

6 See International Organization for Migration, (Geneva: International Organization for Migration, 2019).
7 Andrew Wasike, "World in Progress: Somalia's Climate Refugees," *Deutsche Welle* (September 18, 2019), url: www.dw.com/en/world-in-progress-somalias-climate-refugees/av-50480611.
8 Wasike, "World in Progress."
9 Rebecca Ratcliffe, "Two Million People at Risk of Starvation as Drought Returns to Somalia," (June 6, 2019), url: www.theguardian.com/global-development/2019/jun/06/two-million-people-at-risk-of-starvation-as-drought-returns-to-somalia.
10 See International Committee of the Red Cross, "A Drought So Severe It Has a Name," *International Committee of the Red Cross* (September 26, 2019), url: www.icrc.org/en/document/somalia-conflict-drought-so-severe-it-has-names.
11 See Mathew E. Hauer, Jason M. Evans and Deepak R. Mishra, "Millions Projected to be at Risk from Sea-Level Rise in the Continental United States," *Nature Climate Change* 6 (2016): 691–695.
12 William Arms Fisher (composer) and Katharine Lee Bates, *America the Beautiful* (Boston: Oliver Ditson Company, 1917).
13 See Hauer, Evans and Mishra, "Millions Projected to be at Risk," 691–695.
14 See Thom Brooks, "Climate Change Justice," *PS: Political Science and Politics* 46 (2013): 9–12.
15 Intergovernmental Panel on Climate Change (IPCC), "Special Report on the Ocean and Cryosphere in a Changing Climate: Summary for Policymakers" (2019), 5, url: www.ipcc.ch/site/assets/uploads/sites/3/2019/11/03_SROCC_SPM_FINAL.pdf.
16 IPCC, "Special Report on the Ocean and Cryosphere," 6.
17 IPCC, "Special Report on the Ocean and Cryosphere," 11.
18 See The Royal Society, "How Fast Is Sea Level Rising," (March 2020), url:https://royalsociety.org/topics-policy/projects/climate-change-evidence-causes/question-14/?gclid=EAIaIQobChMIqsC5hc7m6AIVE-3tCh3r2A4-EAAYASAAEgJQR_D_BwE.
19 See IPCC, "Special Report on the Ocean and Cryosphere," 9.
20 See IPCC, "Special Report on the Ocean and Cryosphere," 12.
21 Intergovernmental Panel on Climate Change (IPCC), *Climate Change 2014: Impacts, Adaptation and Vulnerability. Part A: Global and Sectoral Aspects* (Cambridge: Cambridge University Press, 2014): 4 url: www.ipcc.ch/site/assets/uploads/2018/02/WGIIAR5-PartA_FINAL.pdf and IPCC, "2014: Summary for Policymakers," in *Climate Change 2014: Impacts, Adaptation, and Vulnerability* (Cambridge: Cambridge University Press): 4–8 (available on the IPCC website: http://ipcc-wg2.gov/AR5/images/uploads/WG2AR5_SPM_FINAL.pdf.).
22 See Tim Woollings, "How Climate Change Could Be Affecting the Jet Stream," *The Independent* (November 21, 2019), url: www.independent.co.uk/news/science/climate-change-crisis-latest-jet-stream-extreme-weather-a9208901.html.
23 IPCC, *Climate Change 2014*, 4–6. On environmental refugees, see Sujatha Byravan and Sudhir Chella Rajan, "The Ethical Implications of Sea-Level Rise Due to Climate Change," *Ethics and International Affairs* 24 (2010): 239–260; Avery Kolers, "Floating Provisos and Sinking Islands," *Journal of Applied Philosophy* 29 (2012): 333–343; Cara Nine, "Ecological Refugees, States Borders, and the Lockean Proviso," *Journal of Applied Philosophy* 27 (2010): 359–375; and Mathias Risse, "The Right to Relocation: Disappearing Island Nations and Common Ownership of the Earth," *Ethics and International Affairs* 23 (2009):

281–300. On immigration more generally, see Thom Brooks, *Becoming British: UK Citizenship Examined* (London: Biteback, 2016).

24 See National Geographic, "Great Pacific Garbage Patch," *National Geographic*, url: www.nationalgeographic.org/encyclopedia/great-pacific-garbage-patch/.

25 See World Health Organization, "Coronavirus Disease (COVID-19) Pandemic" (2020), url: www.who.int/emergencies/diseases/novel-coronavirus-2019.

26 See Nick Paton Walsh and Vasco Cotovio, "Bats are Not to Blame for Coronavirus: Humans are," *CNN* (March 20, 2020), url: https://edition.cnn.com/2020/03/19/health/coronavirus-human-actions-intl/index.html?iid=ob_article_organic sidebar_expansion and David Cyranoski, "Mystery Deepens over Animal Source of Coronavirus," *Nature* (February 26, 2020), url: www.nature.com/articles/d41586-020-00548-w.

27 See NASA, "Overview: Weather, Global Warming and Climate Change" (2020), url: https://climate.nasa.gov/resources/global-warming-vs-climate-change/.

28 See Michael D. Mastrandrea and Stephen H. Schneider, *Preparing for Climate Change* (Cambridge: MIT Press, 2010): 22–23.

29 David A. Weisbach, "Climate Policy and Self-Interest," in Stephen M. Gardiner and David A. Weisbach (eds.), *Debating Climate Ethics* (Oxford: Oxford University Press, 2016): 170–200, at 174.

30 See Chris Mooney, "How Climate Change Could Counterintuitively Feed Winter Storms," *Washington Post* (January 4, 2018), url: www.washingtonpost.com/news/energy-environment/wp/2018/01/04/how-climate-change-could-counterintuitively-feed-some-winter-storms/.

31 See U.S. President Donald J. Trump's December 29, 2017 tweet sent at 1.01am ("In the East, it could be the COLDEST New Year's Eve on record. Perhaps we could use a little bit of that good old Global Warming that our Country, but not other countries, was going to pay TRILLIONS OF DOLLARS to protect against. Bundle up!"), url: https://twitter.com/realDonaldTrump/status/946531657229701120. More water in the atmosphere freed from melting ice caps and glaciers give extreme weather a greater potency and frequency as the added water can transform an average weather event to anything-but-normal, including snow storms. If it needs saying, cold weather in one location does not disprove or contradict the fact of a rising global temperature taking all locations together into account.

32 Sanitation is a hugely important issue frequently overlooked. Sewage treatment and waste disposal are essential areas we must address – especially with a growing global population on a fixed-sized planet. I explore this issue in Chapter 3.

33 The Intergovernmental Panel on Climate Change (IPCC) website is: www.ipcc.ch/.

34 See Peter T. Doran and Maggie Kendall Zimmerman, "Examining the Scientific Consensus on Climate Change," *EOS* 90 (2009): 286–300. For an exception, see Christopher Booker, *The Real Global Warming Disaster* (London: Continuum, 2009). Nonhuman animals make a contribution, but they are not the primary contributor. See NASA, "The Causes of Climate Change" (2020), url: https://climate.nasa.gov/causes/.

35 See Peter Singer, *One World: The Ethics of Globalization*, 2nd ed. (New Haven, CT: Yale University Press, 2004): 16. There are many different greenhouse gasses beyond carbon emissions, including aerosols and methane. While methane is more potent, it remains in the atmosphere for only about 12 years whereas carbon emissions are created in far larger quantities by human activity and they may remain up to 200 years in the atmosphere. See Stephen Gardiner, "Ethics and Global Climate Change," *Ethics* 114 (2004): 555–600, at 561.

36 See Intergovernmental Panel on Climate Change (IPCC), *Special Report: Global Warming of 1.5°C* (Geneva: IPCC, 2020): 51.
37 Intergovernmental Panel on Climate Change (IPCC), *Climate Change 2013: The Physical Science Basis* (Cambridge: Cambridge University Press, 2014): 472–473.
38 See Jonathan Shell, "Nature ad Value," in Akeel Bilgrami (ed.), *Nature and Value* (New York: Columbia University Press, 2020): 1–12, at p. 3.
39 See Mastrandrea and Schneider, *Preparing for Climate Change*, 8.
40 See Mastrandrea and Schneider, *Preparing for Climate Change*, 12.
41 The United Nations Framework Convention on Climate Change website is: https://unfccc.int/.
42 United Nations Framework Convention on Climate Change, Articles 2 and 31.
43 See UN Environment Programme, "Facts about Climate Emergency" (2020), url:www.unenvironment.org/explore-topics/climate-change/facts-about-climate-emergency.
44 For example, see Albert Gore, *Earth in the Balance: Ecology and the Human Spirit* (Boston: Houghton Mifflin, 1992); David A. King, "Climate Change Science: Adapt, Mitigate, or Ignore?" *Science* 9 (2004): 176–177; and Martin L. Parry, Nigel W. Arnell, Anthony J. McMichael, Robert J. Nicholls, Pim Martens, R. Sari Kovats, Matthew T. J. Livermoore, Cynthia Rosenzweig, Ana Iglesias and Gunther Fischer, "Millions at Risk: Defining Critical Climate Change Threats and Targets," *Global Environmental Change–Human and Policy Dimensions* (2001): 181–183.
45 See His Royal Highness Prince Charles, the Prince of Wales, "Sustainability," (2020), url: www.princeofwales.gov.uk/sustainability. Prince Charles's inspiring efforts have been recognized by his winning a Global Environmental Citizen Award from the Harvard Medical School Center for Health and the Global Environment in 2007. Other recipients include Kofi Annan, Jane Goodall, Al Gore and Alice Waters.
46 There are some noteworthy criticisms of IPCC findings. See Ronald Bailey (ed.), *The True State of the Planet: Ten of the World's Premier Environmental Researchers in a Major Challenge to the Environmental Movement* (New York: Free Press, 1995); Bjorn Lomborg, *The Skeptical Environmentalist: Measuring the Real State of the World* (Cambridge: Cambridge University Press, 1998); and Bjorn Lomborg, *Cool It: The Skeptical Environmentalist's Guide to Global Warming* (New York: Vintage, 2008).
47 My analysis accepts the global scientific consensus about climate change and its effects. This section has made use of free online resources so that the reader can examine these findings further in greater detail. But no such detail is necessary to engage with my subsequent critique of existing proposals and recommended further actions.
48 See Michael Bloomberg and Carl Pope, *Climate of Hope: How Cities, Businesses and Citizens Can Save the Planet* (New York: St Martin's Griffin, 2018); Mayer Hillman, *How We Can Save the Planet: Preventing Global Climatic Catastrophe* (New York: St Martin's Press, 2008) and Nicholas Stern, *A Blueprint for a Safer Planet* (London: Bodley Head, 2009). While I cite these examples, I do admire each of these works. For a perspective arguing against "the 'problem-solving' framing of climate change," see Mike Hulme, *Why We Disagree about Climate Change: Understanding Controversy, Inaction and Opportunity* (Cambridge: Cambridge University Press, 2009): 328–329.

49 Mitigation and technologically-driven adaptation interrelate. For example, the UN Environment Programme understands mitigation as "efforts to reduce or prevent emissions of greenhouse gases. Mitigation can mean using new technologies and renewable energies, making older equipment more energy efficient, or changing management practices or consumer behaviour" (UN Environment Programme, "Mitigation" (2020), url: www.unep.org/explore-topics/climate-change/what-we-do/mitigation). See also James Lovelock, *The Ages of Gaia: A Biography of Our Living Earth*, 2nd ed. (Oxford: Oxford University Press, 2000).
50 See Darrel Moellendorf, "Treaty Norms and Climate Change Mitigation," *Ethics and International Affairs* 23 (2009): 247–265.
51 See Anthony Giddens, *The Politics of Climate Change* (Cambridge: Polity, 2009): 13.
52 See Stephen M. Gardiner, *A Perfect Moral Storm: The Ethical Tragedy of Climate Change* (Oxford: Oxford University Press, 2011).
53 I discuss end-state solutions – and why they are a problem – at greater length in Chapter 4.
54 See Thom Brooks, "How Not to Save the Planet," *Ethics, Policy and Environment* 19(2) (2016): 119–135.

1 Mitigation

The ecological footprint

"The Earth is the mother of all people, and all people should have equal rights upon it."

– *Chief Joseph*

Mitigation as a broad tent approach

Climate change is already happening. We cannot choose how we might avoid this scientific fact and its consequences. It's too late for that. However, this does not mean we lack choices: we have no excuse to do nothing. Indeed, there is much that we can do. But is any available option able to prevent any further climate change? – these first three chapters address this question from different perspectives that claim we can.

The most common approach to thinking about climate change ethics is *mitigation*. This approach is about reducing our environmental impact. We are changing our climate already through human activities like burning fossil fuels. Mitigation approaches seek to prevent further changes through conservationism, such as significantly reducing greenhouse gas emissions. However, pro-mitigation advocates differ on how to bring such changes about.

Mitigation approaches encompass a wide tent covering a diversity of policy proposals. I will examine the two most popular proposals: this chapter will consider the idea of the ecological footprint and the following chapter will look at the polluter pays principle. It is not my claim that all pro-mitigation (and pro-conservationist) advocates endorse, or should endorse, both proposals. Most of these advocates do support one or the other. My aim is to show how each such policy proposals aims to provide a solution to the problem of climate change – and to highlight the difficulties each faces in attempting to achieve its aims.

To summarize, we can accept the global scientific consensus that the problem of climate change is real, caused by human activity and leads to several significant consequences ranging from threatening coastal communities and more severe weather patterns to drought and risk of extinction for plant and animal species. But my point in this and following chapters is that the proposed "solutions" do not altogether solve this very real and grave problem. We will later consider an alternative approach that can help us make progress.

The ecological footprint

One of the most popular and powerful pro-mitigation proposals is the idea of an *ecological footprint*.[1] This footprint is a measure of our planet's human carrying capacity: it is the maximum amount of resource consumption that can be sustained by every individual indefinitely.[2] It is best described by Mathis Wackernagel and William E. Rees:

> The Ecological Footprint concept is simple, yet potentially comprehensive. . . . It is about humanity's continuing dependence on nature and what we can do to secure Earth's capacity to support a humane existence for all in the future. Understanding our ecological constraints will make our sustainability strategies more effective and livable.[3]

The idea is that establishing the size of our ecological footprint is critically important for making concrete our full impact on nature – and when all such impacts from every individual is added together it makes clear that such impact is unsustainable. The ecological footprint clarifies this problem and provides the consumption threshold that we should aim to fall under if we desire a sustainable forevermore.

The ecological footprint provides us with a powerful tool for making us more aware how damaging our impact on the planet can be serving a pedagogical function. Climate change is a global phenomenon. It is a problem caused by human activities everywhere. It can be very difficult for even the more environmentally aware to understand how everyday routines taken for granted add up to an unsustainable lifestyle. By measuring our impact in terms of a footprint, we can see how our local actions have global implications. "Think locally, act globally" has been an effective message for many decades in raising awareness of our environmental impact. The footprint demarcates a safe space, but only if we live within its strict boundaries.

Every individual has an ecological footprint. Our footprint is not separate from ourselves because we are not separate from the environment. We eat, drink and breathe the natural world – and our consumption habits have consequences for it.[4] Nor could we survive independently from nature. Our sustenance and wastes do not come and go from nowhere, but a somewhere in our natural world.[5]

The ecological footprint is about more than what share of the land to sustain us, but rather the proportion of the ecosystem, incorporating the bioproductive land and sea required.[6] Somewhat confusingly, carving up a sustainable slice of the Earth's ecosystem is not the same as drawing lines on a map where we might divide people across equal plots each with the same share of food, water and resources. Instead, our equally sized footprint slices of the ecosystem are a generic, ideal-like space found nowhere in particular. The ecological footprint is a measure and not a place.[7]

Some of us consume more resources than others. The greater an individual's consumption, the larger his or her ecological footprint. A benefit of the ecological footprint is it provides us with a measure of when our consumption of natural resources has grown too large for it to be sustainable indefinitely. If we do not collectively live within the confines of our individual footprint, our total consumption may breach a sustainable level – and we would contribute to climate change and its damaging consequences.

The ecological footprint is *conservationist* in its aims. This is because it will require significant reductions in human consumption, including greenhouse gas emissions and so limiting our reliance on nonrenewable natural resources. Anything less would leave a great many of us where we are – namely, living far beyond our ecological footprint and continue contributing to climate change with its potentially dangerous consequences.

The ecological footprint is also *egalitarian* at its core. Each individual must live within an equally sized ecological footprint. All people are treated equally. This perspective informs how we determine exactly how large we should establish our footprints. We consider what size could be equal for all while guaranteeing human sustainability. Therefore, we are not permitted to consume and pollute more than others. If we did do so, we would live beyond our sustainable means taken collectively. We share global conservation equally as equal partners with equal shares.[8]

This egalitarian commitment is also seen as an integral part of the ecological footprint's broader appeal. It places everyone on an equal footing in relation to our morally and pragmatically permissible ecological impact. We have equal moral claims to the same ecological space for sustainable survival. No one has a natural right to use or enjoy more than others. The footprint is also pragmatic in setting a global cap that the collective footprints of all individuals across the world cannot breach.

Equal footprints as fair shares

There is a second, related way to determine the size of our ecological footprint. This is to consider what is the correct size of equal, fair shares of the absorption capacity for the atmosphere's sink.[9] This is not an appeal to consider only climate change-related impacts in the sky while neglecting the land. The atmosphere is a source of oxygen making our planet inhabitable, but the wider atmospheric sink refers to a broader capacity to absorb pollutants like greenhouse gas emissions – largely achieved through plants, the ocean and the soil.

There is a limit to how much greenhouse gas emissions our atmospheric sink can absorb safely. For example, drinking water can have trace amounts of elements like lead or sodium while remaining safe for consumption even if, in large quantities, these same elements would render the water dangerous. Burning a wood fire produces emissions that might pose little, if any, harm in small quantities and irreparable damage in large amounts.[10] Our planet is not separate from us. Polluting our air, land and water damages our environment and the wider ecology impacting on animals, fish and plants – and it imposes harms on *us*. If we were to make planet inhospitable for human life, it would be an environmental catastrophe with genocidal consequences. So we must live within the limits of what our atmospheric sink can absorb. Or so the argument goes.

The motivating core ideas from the ecological footprint – of equality and pragmatism – remain. But instead of claiming every individual has an equal ecological footprint, the argument here is about equal rights to shares of the atmospheric sink. For instance, the philosopher Peter Singer, who is the most compelling defender of this view, argues the atmospheric sink belongs to all of us in common. No one has any right to a greater share of this space than anybody else. He says: "The atmosphere's capacity to absorb our gases has become a finite resource on which various parties have competing claims. The problem is to allocate those claims justly."[11]

Restricting every individual to no more than their fair share is enormously important. If you were to use more than your fair share, then this would wrongly deprive me of my share, should we ensure that collectively we do not consume more than the atmosphere can safely and sustainably absorb. If we did not ensure our global consumption was restricted, then those using more than their fair share contribute to a situation far more grave whereby my right to a safe and sustainable future is threatened. This is because the unfairly large consumption of others breaching safe limits for our atmospheric sink's capacity to absorb emissions contributes to further climate change and its damaging consequences.

It is very clear that individuals living in affluent countries are consuming far more than their fair share – and beyond any equally sized ecological footprint. Singer powerfully illustrates this problem:

> The average Americans, by driving a car, eating a diet rich in the products of industrialized farming, keeping cool in summer and warm in winter, and consuming products at a hitherto unknown rate, uses more

than fifteen times as much of the global atmospheric sink as the average Indian.[12]

The problem is easily misunderstood. The example can give the impression that driving and the like are wrongs *in themselves* to be avoided at all costs. But this is not strictly correct. Singer's argument is more nuanced, or so it seems to me. If the atmospheric sink had greater capacity, then the climate's tolerance level would be higher. The size of a "fair share" is relative to the global population and the capacity of the atmospheric sink to absorb any emissions. A smaller population and more expansive sink would allow for a larger-sized fair share than is permissible under current circumstances. And if driving, a rich diet and the like could be used or consumed with less impact through improved technology, then it would be less problematic. End of story.[13]

The problems identified by Singer above are twofold. First, the average American is using a far larger – and unfair – share of the atmospheric sink than others. So the first problem is every individual has equal rights to a fair share, but some are consuming far more than others. Secondly, the amount of these unequal shares contributes to humanity's breaching the atmosphere's total carrying capacity. In other words, far more is being emitted than can be safely and sustainably absorbed.

And it's worth noting that having a right to a fair share does not mean every individual should consume their full share. A fair share is an acceptable limit. If using less was more desirable, this would reduce our environmental impact further with only beneficial effects on the global climate.[14]

These different ways of thinking about equal shares of a space held in common – either as an ecological footprint or as an equal share of the atmospheric sink – lead to the same general view: we need to determine equal shares for each individual that are enough to for our survival but not so much that they threaten to unsustainably damage the environment. This level will require far fewer greenhouse gas emissions than at present.[15] The broad approach is a mitigation strategy aiming at conservation levels permitting a sustainable forever-after.

Living beyond our means

The justification for mitigation strategies generally is that we are living beyond our means – measured either through the use of an ecological footprint or share of the atmospheric sink. Before turning to a critical look at these proposals, it is important to recognize the problem they attempt to solve.

There is much evidence to suggest that the populations in many countries are living beyond the limits of their ecological footprints. For example, consider Singer's comparison of average American consumption versus average Indian consumption. If we measure world carbon emissions between 1950 to 1986, it reveals that "the United States, with about 5 percent of the world's population at that time, was responsible for 30 percent of the cumulative emissions, whereas India, with 17 percent of the world's population, was responsible for less than 2 percent of the emissions."[16]

This *fair share gap* between affluent countries and the developing world mirrors disparities in relative wealth, too. The United States is responsible for 30 percent of the cumulative emissions and 30 percent of the world's wealth.[17] Similarly, India is responsible for less than two percent of emissions and has less than four percent of global wealth.[18] Unequal consumption is mirrored in unequal wealth. Those with the most wealth use more than their fair share of resources – and the fair share gap has created an environmentally unsustainable situation.

The fair share gap has wide-ranging consequences. Consuming more than our fair share can contribute to our having responsibility for harming, or threatening harm, to others.[19] The idea of harm is a complex subject that we will examine more closely in Chapter 2. Nevertheless, it is important to make clear here that the core problem with using more than a fair share is that this can contribute to environmental *damage*, not merely environmental *change*. Plant and animal species may adapt and evolve because of environmental changes – whether or not the result of human activities – and this does not confirm environmental damage.[20]

The conceptual difference that I want to draw between damage and change is what we might call the difference of *damaging environmental change* versus *mere environmental change*. This is an important distinction.

Species have evolved since their first appearance – including long before human beings existed. This kind of evolution is a result of nonhuman-caused environmental factors that we might identify as *mere* environmental change. Some of these events were significant, such as the mass extinction associated with the Cretaceous period 66 million years ago ending the dominance of dinosaurs on the planet. As this example exemplifies, such changes can have deadly consequences but are lacking in any cause based on the activities of responsible agents. The cause of change is *mere nature*.[21] Such events are "natural" insofar they are not a product of human activities.[22]

In contrast, we might define *damaging* environmental change as changes detrimental to plant and animal species caused by human activity. Such changes endanger the continued flourishing of the nonhuman world, including raising the risk of extinction, on account of human responsibility.[23] So the difference is not in the level of danger, but its *causal origins*. Mere change can lead to extinction, but it is not a consequence of responsible agents.

The potential value of this distinction – which we will revisit in Chapter 4 – is clarifying the wrongs of problems like the fair share gap. The problem is not that mere environmental changes might occur, but rather that using more than my fair share will cause environmental change that have a negative global consequence. A naturally occurring death or loss is a tragedy, but when this is caused it can become a wrong.[24] Similarly, proponents of mitigation see breaches as human-caused climate changes that are wrongful – and widespread.

Five problems

The ideas of the ecological footprint and fair shares of the atmospheric sink are attempts at "solving" the problem of climate change. If only we confined ourselves to the same-sized footprint or share, a happy sustainable ever-after is within our grasp. There are five key problems with this view. These will be set out in this section:

Anthropocentrism

The first problem is that it is an *anthropocentric* approach.[25] Consider how we determine the size of a footprint or fair share. These are both calculated in terms of human consumption of the natural world. We measure the impact of our activities upon the environment – and we do not have the same regard to the environmental impact of plant and animal species. What matters is finding a sustainable footprint or fair share to ensure a sustainable future for human beings.

This anthropocentric perspective may be helpful in thinking about how to plan a sustainable future for humanity. One issue is that whatever the share of ecological space suitable for our flourishing might be very different from what amount would best sustain plant and animal species more broadly.[26] So while the ecological footprint may lead to a future of sustainable human activities, this may not include continued flourishing of the natural world. If human sustainability can be guaranteed at the cost of nature's diminished flourishing, then the ecological footprint is an approach that might better ensure the conservation of human beings at the cost of conserving nature.[27]

Such a result would have the paradoxical consequence of our "green" approach recommending a potentially much less green future. This might appear especially problematic when we consider that human flourishing – at least on the account of those who argue for this approach – is not wholly separable from the flourishing of the natural world which is our context. And yet its commitments to anthropocentrism create this separation.

Fairness and inequality

The ecological footprint and fair shares are thought to treat all persons on an equal and fair footing.[28] Yet, fairness and equality are in deep tension. If this criticism is correct, it undermines central motivations behind why many support these approaches. Let me explain.

Consider the following. We are each to live within the bounds of our ecological footprint to ensure a sustainable future and this footprint is uniformly equal for all. One problem with this view is equal footprints may be unfair – and so fairness demand unequally sized shares.

The reason behind this justified inequality is because my fair use of the environment may involve different amounts of necessary nourishment and other bodily needs than others. The beginning and end of life – childbirth and old age – may require greater resource needs and so the need for a larger footprint than others who lack similar needs. There are also potential gender differences in resource needs pertaining to pregnancy. So there is no "one size fits all" ecological footprint we may apply to everyone – and nor is there any single, fixed footprint for any individual during our lifetime. Our footprints change over our lives *as we change*. Restricting every individual to the same-sized footprint fails to recognize the differences between individuals.

But there is more. Suppose we were able to account for these individual differences between persons over a lifetime, such as to argue equal footprint should be then be based on an average accounting for fluctuations in resource needs over time. We should still argue a fair solution is to avoid agreeing to the same-sized footprints or shares. This is because even if we could account for average differences between genders and over a lifetime, our resource needs will necessarily vary in relation to the local climate we inhabit. The energy needs of living through four seasons like in New England with humid summers and snowy winters is very different from the dry heat of the Arizona sun.[29] Or put another way: the resource needs of living in the Arctic Circle or mountain ranges will be different from the needs of living in a jungle or in a desert all things considered. Each area has very different climates – and correspondingly resource needs – than the others.

Determining fair ecological footprints may often entail unequal sizes to accommodate these climatic differences. The point is the footprints that restrict us must work for the local climate we occupy. These climates will differ – and so too our footprints if we are to be treated fairly on an equal footing.

Restricting every individual to no more than the same-sized footprint or share treats them unequally. This is because the same footprint does not correspond to the same relative – or even necessary – resource needs that individuals will have. If we were to insist on footprints or shares in relation to these needs, situating everyone equally will lead to unequal footprints. In other words, individuals of very different natural body types, sizes and metabolisms have different resource needs relative to their having the same level of nourishment and flourishing. Equal treatment should then entail different (and unequal) footprints.

Equality and unfairness

Equal ecological footprints may also be unfair. For example, societies have developed differently in relation to their wealth and technological advances. This can raise significant and long-term problems.

Suppose we agreed to an average-sized ecological footprint or share. Each country was restricted to a total footprint equal to the relative size of its relevant population. A country with twice as many citizens as another would have double the footprint. However, each country would have the exact same footprint per capita.

This leads to the problem of equality creating unfairness. An illustration of this is that an equal per capital footprint would ossify – and render more permanent – the relative global positions of the more affluent and technologically advanced countries in contrast to developing countries which lack such infrastructure. The former would be in a much better position to make the most from their limited footprint.[30] This would permit them to better retain their position of global privilege over less capable societies.[31]

The ecological footprint would not treat persons equally or fairly while engendering a more unequal global distribution of wealth. It might be argued that creating an unjust world where affluent states could secure more permanently their relative economic dominance over developing countries could be a price worth paying if this ensured a sustainable ever-after.[32]

In reply, there are two flaws to this powerful objection. The first is that equality and fairness are central values of the ecological footprint and fair shares approaches. Their equality and fairness are not mere instrumental values, but a part of its justification and, at least in the eyes of its proponents, a key source of its attractiveness. They claim to show a path to saving the world from climate change's dangerous consequences by treating us equally and fairly. So if they led to ossifying global structural inequality and unfairness, this would be fatal to their central values. The second flaw is they do not "solve" the problem so the price paid would be in vain – this argument will be explored in Chapter 4.

Should the same emissions by different groups be treated similarly?

The ecological footprint and fair shares approaches treat emissions similarly. We consider how much the atmospheric sink can sustainably absorb and divide shares equally based on the size of the current population.

Henry Shue makes an influential case that not all emissions should be treated the same. This is not an argument about how carbon emissions and methane have differential impacts on the environment although that is true. Instead, it is a matter of justice. Shue argues we have an inalienable right to emissions required for our survival or at some minimum threshold of quality of life.[33] (This presupposes that ensuring a minimum threshold for all is possible without contributing further to unstainable levels of climate change.) Since these emissions are required for our survival, they are inalienable – they cannot be traded or appropriated by others. While there may be scope for minor fluctuations in the exact size any individual has a right to, these inalienable *subsistence emissions* have an equal status for every individual.

In contrast, Shue claims that emissions associated with what he calls luxury goods have a different status. Not only do we lack the same inalienable right to these kinds of emissions, we should avoid their creation in our efforts to tackle climate change.

Shue asks the question of whether the same kinds of emissions created by different grounds should be treated similarly. His answer is no. Consider a hypothetical example. Both the affluent state of *Richland* and the developing state of *Pooria* produce carbon emissions.[34] The difference is that most of the Richland's emissions are created to produce luxuries for their rich citizens while Pooria might be more for meeting some minimal threshold for its poor citizenry. While each burns fossil fuels, Pooria is more justified in doing so – and even if its total emissions was roughly similar to the Richland. The *justification* for creating emissions is more important than its *amount*. This point is expressed well by Paul Harris, who says: "One thing that seems unassailable from the perspective of world ethics and global justice is that greenhouse gas emissions required for subsistence take priority over other kinds of emissions, and ought not be subject to any kind of limit. . . . Arguing the point is as good as saying that Rwandans should die so that some Virgin Islanders can recharge their mobile phones."[35]

There are two concerns with this influential proposal. The first is everything rests on our being able to clearly and consistently distinguish between greenhouse gas emissions that are necessary (or, in Shue's terminology "subsistence emissions") from those that are unnecessary (e.g., emissions related to producing luxuries).[36] It is perhaps in the eye of the beholder what should count as a luxury as the same things can count differently as such in various communities around the globe.

A second concern focuses on the difference between *necessary* and *unnecessary* emissions. If we can distinguish emissions that meet the requirement for necessity, then it is not clear this must be sustainable especially if there were overpopulation. If what counts is justification and not the total amount emitted, a developing country like hypothetical Pooria could emit unsustainably high amounts of greenhouse gases to feed its hungry people using its inefficient, coal-powered infrastructure. However, if what we want is an approach to reducing our emissions, Shue's proposals would undoubtedly, if followed, lead to a significant reduction in emissions today, but these proposals do not require we cannot emit more than the environment might sustainably allow if we could justify its necessity as an inalienable right. We could commit ourselves to irreversible environmental damage fulfilling rights to those present – against the rights of those in future generations having to live with the consequences.

Overpopulation as a tragedy of the commons

An equal footprint or fair share is based on a problematic assumption. This is an article of faith that if we divide equal shares of the ecosystem or atmospheric sink there will be enough to go around and sufficiently enable all individuals to flourish on a sustainable basis. Perhaps it shares a background belief found in the Old Testament that "the Lord will provide" no ifs or buts – we need only ensure that these shares are fairly distributed.[37]

This assumption is problematic. Consider the example of *the tragedy of the commons*.[38] There is a common pasture that can be used for grazing cattle. We may each bring our livestock to feed and we freely share this resource without any restrictions. Since we can earn more from our cattle if we owned more of them, we each bring larger herds to feed on the pasture. The problem is that this situation can quickly escalate to the point where our commonly shared asset is overgrazed leading to our becoming unable to sustain our livestock.

The tragedy of the commons is supposed to illustrate that our individual actions may lead to less than optimal overall outcomes when different people have access to an unregulated good. The situation is a tragedy because the common good, if left unchecked, can be exploited and through our common usage lead to its decimation with catastrophic results.

This hypothetical example is a popular illustration for many about how an unregulated common use of the atmosphere for greenhouse gas emissions can lead to this natural resource becoming poisoned and inhospitable – like a pasture overgrazed by livestock. The lesson we are to learn that if this commonly shared resource was managed properly we might not be able to bring as many cattle to feed (or emit greenhouse gas emissions) as often as we might like, but we can make it sustainable if only we keep to an equal share of the pasture (or footprint).

But this presupposes that there is no overpopulation. It is possible that the number of human beings could increase to the point where an equal share of resources divided between individuals would not be enough to sustain them all indefinitely. This is another case that having equal shares does not guarantee sustainability. It is not so much the problem of our number *per se*, but our overall consumption that is the central issue.

The problem of overpopulation and war

There is a final problem of overpopulation. The size of an ecological footprint or fair share is determined by considering the Earth's sustainable carrying capacity or atmospheric sink – and then dividing this up by the total population. Overpopulation would make it impossible for all to live within an ecological footprint of equal size. This is because their footprints would point beyond sustainability when taken together. Ecological footprints require the absence of overpopulation and assume there is sufficient sustainable ecological space for all – a point that has been made already.

But consider a hypothetical *perverse incentive*, or unintended incentive with an undesirable consequence, to "solve" the problem of overpopulation. The size of every individual footprint is the same and relative to an equal share of the Earth's sustainable carrying capacity. This size will fluctuate relative to the size of the global population. If there are more people, then we carve up a fixed carrying capacity by a greater number – and so leading to smaller individual shares. The only way that our footprints might become bigger is if the global population were less.

Hypothetically, this could give unjust regimes a perverse incentive to attack civilians elsewhere in order to expand the size of their footprints.[39] By reducing the population, this would create extra space to be shared by the remaining population – and so expand the size of everyone's footprints. If an unjust regime wanted to increase the footprint size of its citizens (and coincidentally of all other citizens, too), war could be an unjust means of achieving this end.

It is *not* my view that supporting the use of an ecological footprint would create a military conflict. Any such action would violate international law and almost certainly produce a large amount of emissions as well. It would be self-defeating as both an ecological and humanitarian disaster. However, my point is only to raise a hypothetical incentive that follows from the logic: if fewer people mean larger footprints, there is a perverse incentive that could theoretically flow from that.

Carbon trading

A final mitigation proposal to consider is carbon trading, also sometimes called cap-and-trade programs.[40] The idea is that each country possesses shares in greenhouse gas emissions. Each state may have different shares than others. There are two important factors. First, the total number of shares is equal to some sustainable amount of emissions the atmospheric sink can withstand safely. If every share was used, environmental sustainability is maintained. This is only threatened if states used more than the total shares. Second, states cannot produce more emissions than for what they have shares for. If a country wishes to produce additional emissions above and beyond its originally allocated share, then it may purchase – or trade for cash – emissions credits from others.

This system has several advantages. The first is it sets a clear overall limit, or "cap," on the total permissible emissions. If this restriction is not breached, this would satisfy the conservation-friendly mitigation aims of reducing global emissions below a set threshold. The second advantage is that affluent states are high carbon emission states. The opportunity to purchase credits makes their fossil fuel dependency more expensive to maintain and so provides a cost aimed at disincentivizing this dependency. At the same time, purchasing credits renders it easier for states with higher carbon emissions to bring these more gradually to lower levels.

The main problem is that carbon trading may produce a negative effect on our motivations to conserve.[41] Conservationists believe we should conserve to best address the associated dangers of climate change. Therefore, we should not contribute to climate change. Carbon trading does not ensure that each state becomes more sustainable. Instead, it is a mechanism to better guarantee that the global system taken as a whole may become more sustainable. Sufficiently wealthy citizens who can purchase carbon credits may continue to produce increasing emissions than citizens elsewhere. The wealthy few may enjoy greater resource use and material luxuries at the expense of a majority left with much less.[42]

The problem is that the wealthy few are also the largest polluters. Carbon trading may not offer sufficiently attractive incentives to pollute less. Instead, it might lead to ossifying the global status quo.[43] Indeed, some argue that trade liberalization leads to an outsourcing of pollution from industrialized countries to developing countries.[44]

There is evidence to suggest that carbon trading encourages firms to buy certificates to meet emissions targets – rather than to sell credits.[45] In other words, more choose to emit more than their fair share. This suggests the costs are too low for buying credits because the added costs do not outweigh benefits. If credits were more expensive, this would be greater benefits in selling, rather than buying, credits through less reliance on fossil fuels. But this is not how the system works.

An alternative to carbon trading among countries is to introduce personal trading by individuals. This was briefly considered by the British government under Labour Prime Minister Gordon Brown in high impact areas like household energy use or personal transportation.[46] The idea that individuals could spend, trade or owe an individualized carbon budget as they could with money was once seen as an idea "ahead of its time."[47]

At the time, there were serious concerns about the feasibility of these plans based around its complexity. Since dropped, they have not been reconsidered since. Psychologist Adam Corner observed: "a truly equal division of carbon would reveal the carbon gap between the rich and the poor."[48] Analogous to the international context, the system for personal trading by individuals might be largely a vehicle for the wealthy to continue their lifestyles paying a small tax to the less well-off for the maintenance of their privilege. Again, the problem of failing to sufficiently motivate conservation persists at both the international and individual levels. Moreover, climate change is a product of more than our emissions – however crucially important their reduction is for reducing the effects of climate change. A system only looking at one aspect will not tackle the full problem.

Conclusion

Climate change is happening. The most common approach to thinking about the relevant ethical implications is mitigation. This approach is rooted in reducing our environmental impact through conservation efforts like cutting greenhouse gas emissions. However, its advocates disagree on how to bring such changes about – and mitigation approaches encompass a wide tent.

This chapter examined the influential proposals for an ecological footprint or fair share of the atmospheric sink – and related policies like carbon trading – that aim to provide a "solution" to the problem of climate change grounded in equality and fairness. What this chapter has shown is that these proposals face serious objections as a means to ending climate change and living up to their high ideals. Identifying our footprint or fair share is not enough to convincingly mitigate the real and present dangers posed by climate change.

However, and to foreshadow the argument that I will construct over the next few chapters, the inability of these mitigation strategies as a long-term "solution" to climate change *does not* mean that mitigation is unimportant or should be avoided. Whereas E. F. Schumacher claimed "small is beautiful," we might say that *less is best* when it comes to greenhouse gas emissions.[49] But less is no guarantee of sustainable success. We must look further and deeper.

Notes

1 See Jeroen C. J. M. Van den Bergh and Harmen Verbruggen, "Spatial Sustainability, Trade, Trade, and Indicators: An Evaluation of the 'Ecological Footprint'," *Ecological Economics* 29 (1999): 63–74; Steve Vanderheiden, "Two

Conceptions of Sustainability," *Political Studies* 56 (2008): 435–455; Mathis Wackernagel and William E. Rees, *Our Ecological Footprint: Reducing Human Impact on the Earth* (Gabriola Island: New Society Publishers, 1996); Mathis Wackernagel and William E. Rees, "Perceptual and Structural Barriers to Investing in Natural Capital: Economics from an Ecological Footprint Perspective," *Ecological Economics* 20 (1997): 3–24. The World Wildlife Fund has an online footprint calculator here: http://footprint.wwf.org.uk/.)

2 See William E. Rees, "Ecological Footprints and Appropriated Carrying Capacity: What Urban Economics Leaves Out," *Environment and Urbanization* 4 (1992): 121–130, esp. p. 125. See also Dale Jamieson, *Ethics and the Environment* (Cambridge: Cambridge University Press, 2008): 184 and Arjen Y. Hoekstra and Mesfin M. Mekonnen, "The Water Footprint of Humanity," *Proceedings of the National Academy of Science* 109 (2012): 3232–3237. There is an online ecological footprint calculator here: World Wildlife Fund, "How Big is Your Environmental Footprint?" (2020), url: https://footprint.wwf.org.uk/.

3 See Wackernagel and Rees, *Our Ecological Footprint*, 3.

4 See Wackernagel and Rees, *Our Ecological Footprint*, 7.

5 Sanitation and sewage are significant issues rarely touched on in most work in environmental political theory. This is a significant mistake given their clear importance. A sustainable future must be about much more than what we consume in food and water, but also all by-products of this consumption. I discuss this at greater length in Chapter 3.

6 See Mathis Wackernagel, "Methodological Advancements in Footprint Analysis," *Ecological Economics* (2009): 1925–1927.

7 I write "somewhat confusingly" because there seems some confusion among the general public about the kind of space that an ecological footprint embodies. The above clarifications aim at addressing this misconception.

8 This idea is related to the idea of ecological debt. We may owe others an ecological debt whenever we live beyond our ecological footprint. See Simms, *Ecological Debt*, 88.

9 See Singer, *One World*, 28 and Martino Traxler, "Fair Chore Division for Climate Change," *Social Theory and Practice* 28 (2002): 101–134.

10 For example, see Environmental Protection UK, "Using Wood and Coal for Home Heating," (2020), url: www.environmental-protection.org.uk/policy-areas/air-quality/air-pollution-law-and-policy/using-wood-and-coal-for-home-heating/.

11 Singer, *One World*, 29.

12 Singer, *One World*, 31.

13 British readers can replace the expression "end of story" with "full stop." As an American and British dual national, I try to accommodate expressions for both regions.

14 My language is deliberate. Some climate change could have beneficial local effects (for humans) like making once frozen ground agrigable, but would have negative global effects.

15 There is a difference between ecological footprints and equal shares in the global atmospheric sink, strictly speaking. Ecological footprints are a wider measure of impact on global ecology, while equal shares of the atmospheric sink is focused particularly on atmospheric gases, especially carbon emissions.

16 Singer, *One World*, 32.

17 See Anthony Shorrocks, Jim Davies and Rodrigo Lluberas, *Global Wealth Report 2019* (Geneva: Credit Suisse, 2019).

18 Shorrocks, Davis and Lluberas, *Global Wealth Report 2019*.

19 This is one important motivation for many conservationists, but not the only motivation. See Singer, *One World*, 14–50.

20 It is not the presence of carbon emissions or other greenhouse gases that is the problem. Indeed, there is already "a purely natural greenhouse effect" responsible for a warmer surface temperature. John Houghton argues that the Earth's surface temperature is 15°C instead of −6°C. See Gardiner, "Ethics and Global Climate Change," 557–558 and John Houghton, *Global Warming: The Complete Briefing*, 2nd ed. (Cambridge: Cambridge University Press, 1997): 11–12.

21 The Cretaceous period's mass extinction was caused by natural events. Some claim evidence that an asteroid striking Mexico is to blame. Either way, a mere environmental change is a purely natural environmental event that is not caused by the activities of human beings or any other responsible agent. See William J. Broad and Kenneth Chang, "Fossil Site Reveals Day That Meteor Hit Earth and, Maybe, Wiped Out Dinosaurs," *New York Times* (March 29, 2019), url: www.nytimes.com/2019/03/29/science/dinosaurs-extinction-asteroid.html.

22 Of course, human beings are a part of nature. I am not trying to argue for the separability of humans from their environment. But I do want to distinguish between events that would happen "naturally" and those that happen because of human activities and impact.

23 I do not suggest that this is the only nor the primary understanding of harm held by mitigation proponents and other conservationists. Other understandings might include any damage to our aesthetic experience of nature. See Jamieson, *Ethics and the Environment*, 158–162.

24 My use of language is intentional. Not all deaths caused by an individual's actions are normally considered immoral or unlawful, such as where it happens in self-defense. Instead, it is better to say causing death "can" make it wrong such as where it is a case of manslaughter or murder.

25 On anthropocentrism and climate change ethics more generally, see Nicole Hassoun, "The Anthropocentric Advantage? Environmental Ethics and Climate Change Policy," *Critical Review of International Social and Political Philosophy* 14 (2011): 235–257.

26 It might be argued that a view favoring the flourishing of all organic life raises serious problems beyond how we value human beings in relation to animals and plants, but about bacteria and viruses. Nonetheless, a purely anthropocentric view separates human beings in an artificial way from their lived environments. It is this artificiality that is at issue. Anthropocentrism is not objectionable because it prioritizes human beings, but because it considers human beings as separable from the environment. What we might call a *people prioritarianism* could prioritize the interests of human beings but without failing to acknowledge other interests from the natural world. It would very helpful to see anthropocentric views challenged by such people prioritarianism.

27 This position does not deny that we are "trustees of the planet." (See Robin Attfield, *The Ethics of the Global Environment* (Edinburgh: Edinburgh University Press, 1999): 45.) Moreover, we may prefer to favor ourselves over the natural world. My claim is not that this is a view we should reject, but rather that the ecological footprint treats the natural environment as secondary, or perhaps as instrumental, to human sustainability. My thanks to Matthew Noah Smith for raising this issue.

28 See Anil Agarwal and Sunita Narain, *Global Warming in an Unequal World: A Case of Environmental Colonialism* (New Delhi: Centre for Science and Environment, 1991); Paul Baer, "Equity, Greenhouse Gas Emissions, and Global Common Resources," in Stephen H. Schneider, Armin Rosencranz, and John O. Niles (eds.), *Climate Change Policy: A Survey* (Washington, DC: Island Press, 2002): 393–408; and Dale Jamieson, "The Epistemology of Climate Change: Some Morals for Managers," *Society and Natural Resources* 4 (1991): 319–329.

29 As a native of Connecticut and former Arizona resident, this example is based on my personal experience. The former has hurricanes and snowstorms, the latter mostly blue skies and little rain. The resources required to thrive in one are different from the other.

30 See Tim Hayward, *Constitutional Environmental Rights* (Oxford: Oxford University Press, 2005): 198.

31 See Vanderheiden, "Two Conceptions of Sustainability," 446–447. On global justice issues more generally, see Thom Brooks (ed.), *The Global Justice Reader* (Oxford: Blackwell, 2008); Thom Brooks (ed.), *Global Justice and International Affairs* (Boston: Brill, 2012); Thom Brooks (ed.), *Justice and the Capabilities Approach* (Aldershot: Ashgate, 2012); Thom Brooks (ed.), *New Waves in Global Justice* (Basingstoke: Palgrave Macmillan, 2014); Thom Brooks (ed.), *Current Controversies in Political Philosophy* (London: Routledge, 2015) and Thom Brooks (ed.), *The Oxford Handbook of Global Justice* (Oxford: Oxford University Press, 2020). More generally, see Thom Brooks, "Global Justice," in Duncan Pritchard (ed.), *What Is This Thing Called Philosophy?* (London: Routledge, 2016): 68–80.

32 I am indebted to Fabian Freyenhagen for raising this objection.

33 See Henry Shue, "Subsistence Emissions and Luxury Emissions," *Law and Policy* 15 (1993): 39–59.

34 I will return later to using this hypothetical comparison of wealthy "Richland" versus developing "Pooria." There is no specific country in mind behind either fictional state.

35 Paul G. Harris, *World Ethics and Climate Change: From International to Global Justice* (Edinburgh: Edinburgh University Press, 2010): 131–132.

36 See Gardiner, *A Perfect Moral Storm*, 425.

37 See Genesis 22:13.

38 See Garrett Hardin, "The Tragedy of the Commons," *Science* 162 (1968): 1243–1248 and Christopher Knapp, "Tragedies without Commons," *Public Affairs Quarterly* 25 (2011): 81–94.

39 This hypothetical arose out of a discussion with my former colleague Graham Long. He was exploring how, if at all, a just war could ever have climate change as a cause. It occurred to me at the time that climate change could be a cause – but of a decidedly *unjust* war. I harbor strong doubts about just war in general.

40 See Simon Caney and Cameron Hepburn, "Carbon Trading: Unethical, Unjust and Ineffective?" *Philosophy* 69 (2011): 201–234; Cameron Hepburn, "Carbon Trading: A Review of the Kyoto Mechanisms," *Annual Review of Environmental Resources* 32 (2007): 375–393; Cameron Hepburn and Nicholas Stern, "The Global Deal on Climate Change," in Dieter Helm and Cameron Hepburn (eds.), *The Economics and Politics of Climate Change* (Oxford: Oxford University Press, 2009): 36–57, esp. 49–53; J. Kurtzman, "The Low Carbon Diet," *Foreign Policy* 88 (2009): 114–122; M. Lazarowicz, *Global Carbon Trading: A Framework for Reducing Emissions* (London: TSO, 2009); Edward A. Page, "Cosmopolitanism,

Climate Change, and Greenhouse Gas Emissions Trading," *International Theory* 3 (2011): 37–69; Edward A. Page, "Cashing in on Climate Change: Political Theory and Global Emissions Trading," *Critical Review of International Social and Political Philosophy* 14 (2011): 1–15; C. M. Rose, "Expanding the Choices for the Global Commons: Comparing Newfangled Tradable Allowance Schemes to Old-Fashioned Common Property Regimes," *Duke Environmental Law and Policy Forum* 10 (2000): 45–72, at 52–68; Mark Sagoff, "Controlling Global Climate: The Debate Over Pollution Trading," in V. V. Gehring and W. A. Galston (eds.), *Philosophical Dimensions of Public Policy* (London: Transaction Publishers, 2002): 311–318; R. N. Stavins, "Addressing Climate Change with a Comprehensive US Cap-and-Trade System," *Oxford Review of Economic Policy* 24 (2008): 298–321; and T. Tietenberg, *Emissions Trading: Principles and Practice,* 2nd ed. (Washington, DC: Resources for the Future, 2006): 25–47, 192–203. While carbon trading can be understood as part of a mitigation approach, I do not argue or recommend that all conservationists are or should be supporters of carbon trading. I noted at the beginning of this section that conservationism is a large tent encompassing a diverse variety of proposals: carbon trading is part of this diversity.

41 See Andrew Dobson, *Citizenship and the Environment* (Oxford: Oxford University Press, 2003): 2–3 and Eric Posner and Cass Sunstein, "Should Greenhouse Gas Permits Be Allocated on a Per Capita Basis?" *California Law Review* 97 (2009): 51–93.

42 See Shue, "Subsistence Emissions and Luxury Emissions," 39–59.

43 My view is similar to Posner and Sunstein's argument that carbon trading lacks sufficiently attractive incentives for major polluters, such as the United States, to pollute less. My concern goes further: my worry is that not only skepticism about the effectiveness of carbon trading for reducing carbon emissions to the levels required for sustainability, but that the system is likely to return unequal benefits in favor of the affluent and technologically advanced against the more poor and less technologically able.

44 See Werner Antweiler, Brian R. Copeland and Scott Taylor, "Is Free Trade Good for the Environment?" *American Economic Review* 91 (2001): 807–908.

45 For a discussion on this point, see Juan Miguel Rodriguez Lopez, Anita Engels and Lisa Knoll, "Understanding Carbon Trading: Effects of Delegating CO_2 Responsibility on Organizations' Trading Behaviour," *Climate Policy* 17 (2017): 346–360.

46 See Tina Fawcett, "Personal Carbon Trading: Is Now the Right Time?" *Carbon Management* 3 (2012): 283–291.

47 See Environmental Audit Committee, *Personal Carbon Trading, Fifth Report of Session 2007–08 (HC 565)* (London: House of Commons, 2008).

48 Adam Corner, "Personal Carbon Allowances: A 'Big Idea That Never Took Off'," *The Guardian* (April 30, 2012), url: www.theguardian.com/sustainable-business/personal-carbon-allowances-budgets.

49 E. F. Schumacher, *Small Is Beautiful* (New York: Harper & Row, 1973).

2 Mitigation

The polluter pays principle

"Earth provides enough to satisfy every man's need, but not every man's greed."

– *Mahatma Gandhi*

The polluter pays principle

An alternative mitigation strategy to the ecological footprint or fair shares approaches is *the polluter pays principle*.[1] This principle is built on the premise that we have a negative duty to compensate others for the harm we have caused to others. Polluters should compensate others for their damaging greenhouse gas emissions. Their compensation ought to satisfactorily minimize, if not annul, environmental damage relating to these emissions. The polluter pays principle offers us a second means of conservation by reducing emissions and mitigating climate change.[2]

Like the ecological footprint, there is a broad tent of views about how we should understand a polluter pays principle – with different views about which *polluters* should pay, how much these polluters should *pay*, whether such payments can *compensate* adequately for the damage created through pollution, if the higher costs will lead to emission reductions and whether some form of global taxation – such as a resources dividend or border tax adjustment – can bring about global climate justice. This chapter will separate out these interconnected issues to highlight the broad attractiveness of the polluter pays principle for its supporters and to raise serious doubts about its feasibility as a "solution" to the significant problems posed from climate change.

Who should pay and who should be paid?

The polluter pays principle says that those who pollute should pay others because of it. There is much to unpack and consider. An important starting point is identifying *who should pay* and *who receives their payments*. The polluter pays principle suffers from its own version of *an identity problem*. Put simply: we are all the polluters and the victims of pollution. So who pays whom?[3]

One common view is to say the "polluters" are individual countries with collective responsibilities for the emissions produced within their borders.[4] But recall from the Introduction that our carbon and other greenhouse gas emissions can remain in the atmosphere many decades or much longer – and so potentially outliving the individuals who originally emitted them in the first place. This raises issues of whether current generations should provide compensation today for the emissions produced by previous generations. The sins of one generation are paid by the next.[5] We might call this a form of *environmental intergenerational reparations*.

Some advocate that the polluter pays principle can and should be about polluters paying for the full scale of their pollution – even if spread across generations. For example, James Garvey says:

> It is a straightforward fact that some countries have emitted more greenhouse gases – used up more of the planet's sinks, caused more climate change – than others. It's a quantifiable fact: we know something about cumulative emissions.[6]

This *historical* view of the polluter pays principle is attractive for some like Garvey because it seeks to ensure that polluters, as a whole, pay for all pollution created. It takes the principle at its face – and full – value.

But there are real problems with this historical view. The first problem is that our data on cumulative emissions does not go back especially far. Moreover, the further back we go, the less confidence we have with that data. We require a clear rationale for determining both who should pay and how much pollution they should pay for. A historical view is constrained by its too limited data set. This gives rise to a second problem about the relevant cut-off point: how far back should we go to assess past emissions? We cannot go back to the start of human history and our earliest data is neither comprehensive nor as established as more recent data. So with only an arbitrary time period we can do no better, at most, at a polluter pays principle that pays for only that share of historical emissions we have data for. This arbitrary time period is unsatisfactory.

A third problem with a historical view is that many of those who should pay are now dead – and so cannot pay compensation. A policy based on making polluters pay runs flat when they are no longer alive to do so. There is no way to make all polluters pay – and so there is an element of inequity in shifting burdens from the past to the present for a problem created by both.

A related fourth problem is that past generations which have created emissions contributing to climate change were – if we go back far enough – not fully aware or understood the damage that their activities have brought about. Some claim that we cannot hold polluters responsible for activities they did not know caused such harm. While it is arguable their pollution was not made negligently if we went back far enough, then it does appear reckless. Polluting the land, rivers and sky in greater amounts each year did have negative impacts – so this problem that is often raised seems much less persuasive than others.

These problems have made most proponents focus on seeing people living here and now today as the "polluters" who should pay from a starting point in the present.[7] This is at least a more efficient way of handling the problem. The polluters who should pay are the people polluting today only. One limitation of making this restriction is that if the polluters only pay for their own pollution when climate change's effects are happening, at least in part, because of the pollution of past generations, the polluter pays principle has too narrow a focus and cannot fully address the climatic challenge we face. Whatever good it can and should do, the principle will always fall short. Polluters will never pay for all their pollution.

As we have seen above, the polluter pays principle says that those who pollute should pay others because of it. If the polluters to pay are collectively "us," then which others should be paid because of our pollution? One argument is that the compensation paid by polluters should be a payment from affluent countries to the developing world. The affluent are the best able to pay and create the most emissions while the developing world is less wealthy and more vulnerable to the damage that climate change brings.

This view is enshrined in international agreements like the 1997 Kyoto Protocol concerning legal obligations to limit carbon emissions.[8] All states have a shared responsibility, but some are required to do more than others linked to a country's level of economic development. This *common, but differentiated responsibilities* approach recognizes the share commitment all countries have collectively to address climate change and its effects while recognizing some contribute much more (and for much longer) than others. The costs should be distributed unevenly with the biggest and most wealthy polluters compensating the rest. The privileged should support the disadvantaged.

The Kyoto approach has its critics. The most commonly found objection is that if polluters should pay for their pollution what counts is the amount of pollution – and this amount should be treated the same. Consider our hypothetical countries of wealthy Richland and the less affluent Pooria. The objection works like this: if Richland and Pooria each produce the same or similar amount of greenhouse gas emissions, then they should pay the same or similar amount. Their relative economic inequality is a different matter. We should not attempt to twist, or confuse, climate policy with global distributional justice. Pollution is pollution no matter its source.

A different objection is that if polluters should pay, their compensation should only be sent to internationally approved – or perhaps U.N.-administered – projects aimed at addressing the pollution caused by emissions. Otherwise, it is feared payments meant to support the poor based on compensation for contributing to climate changes that may affect them will instead be plundered by elites in those communities.

A final concern about "polluters" is that this is usually taken to mean a political state with fixed borders. This view of the "Westphalian," state-centric international sphere is increasingly outdated. There are overlapping actors including multinational firms and charitable organizations that act across borders – and yet escape inclusion in how we think about global justice that is in urgent need of a refresh. The political state is not the only relevant actor in global affairs – there is a strong case for expanding our conceptual horizon beyond it.[9]

A beneficiary, not polluter, pays principle?

A rival to the polluter pays principle is what is called *the beneficiary pays principle*.[10] This principle accepts the criticism directed towards the historical polluter pays principle which says it is too difficult to know with sufficient provision how much past generations have polluted and which now requires compensation.

So instead of trying to add up a clear view of how much historical emissions that we might have to pay for today, the beneficiary pays principle says that we should total up the benefits that the present generation enjoys off the back of past greenhouse gas emissions. We should pay not for the pollution of past generations, but for the pollution-related benefits that we enjoy today.

This principle is controversial for several reasons. The first is whether there can be much greater clarity on the full benefits we enjoy from past emissions where we lack such a level of precision about these emissions. We do not escape the problem of having inexact data by trying to build a contemporary mode for how we benefited from past emissions using that data.

A second problem is it still holds people today responsible for emissions they did not create, could not prevent, did not ask for and which took place before we were born. And why only disavow the benefits inherited *from emissions*? There are other possible historical wrongs – such as territory or infrastructure – that might have been chosen instead.[11] For these reasons, the beneficiary pays principle appears no more compelling than the historical polluter pays principle it seeks to replace.

Can polluters compensate for damage?

The polluter pays principle entails considerations of harm, compensation and its conservationist potential. Before turning to the issue of how much should be paid linked to harms made possible from climate change, let us now turn to the issue of *compensation* for causing environmental damage.

The polluter pays principle is a means of providing compensation for a wrong. The polluter ought to pay because they should compensate. The notion that polluters should compensate rather than merely pay better invokes the idea that they are addressing a wrong. This highlights the important difference between a *fine* and a *fee*.[12] A fine similarly invokes a wrong whereas a fee does not. This is one reason to understand what the polluter should pay as a fine. Moreover, the idea of a fine might better contribute to a sense of common responsibility relating to climate change.

Compensation might take many forms. There will be winners and losers from breaking an economy free from its dependence on fossil fuels. Some argue that providing social welfare payments alongside increased environmental protections could be a means of compensating firms and their workforce in transitioning to a reduced greenhouse gas emissions future and any negative economic this might bring.[13]

The more prevalent understanding of how the polluter pays principle is instead in compensating others for the harms caused to them. For example, some mitigation proponents argue that we have a duty to use no more than our fair share of the atmospheric sink. If we neglect that duty, we risk breaching the total global emissions that are sustainable and risk harm to others from climate change because we have neglected our duty. We compensate others not because they are in need, but because they have been harmed. We harm others through contributing to climate change and so we owe – as a negative duty and a matter of natural justice – compensation to address our wrongdoing.

However, it is very far from clear whether we can compensate environmental damage.[14] Should it be permissible to provide compensation for making a species extinct? And what if others reject our offer of compensation? The polluter pays principle assumes we can compensate for the environmental damage we cause. It could be argued that the problem is the principle is too strong. Instead of claiming we can and should compensate *in full* for any environmental damage, it should defend our compensating *as best as is possible*, or "partial compensation" (and not "full compensation"). But the general problem does not go away. If environmental goods are noncompensatory goods, then adopting a position of full or partial compensation is inadequate. The principle further assumes that compensation is unproblematic and it would be widely acceptable.

These assumptions presume too much. Environmental goods, such as a species' existence, may not be compensatory goods and we cannot assume all environmental impacts have a discernible monetary cost. Likewise, it remains unclear why we should in principle permit compensation from others to address *our* being subjected to *their* environmental damage.

There is a further issue about whether by increasing the costs of reducing emissions we might also reduce their efficacy, rendering such efforts self-defeating[15] This may be true if some but not all countries signed up to such a principle. Firms wanting to generate emissions more cheaply might move operations elsewhere to a place that lacks a polluter pays principle – and the added revenue stream might incentivize some countries to consider becoming a host for these firms in the short- or even medium-term. Either way, setting the amounts and distributing the compensation is administratively complex.[16]

Finally, the issue of compensation is even more problematic when we consider the complex topic of the potential detrimental effects upon future generations. In brief, the concern is that the individuals making up a future generation could be different depending on climate changes. So to consider compensation or harms to future generations raises deep questions – that I will bracket here – about whether climate change harms a future generation that might not otherwise exist but for the changing climate. The answers are not obvious and a subject of deep philosophical debates.[17]

To summarize: even if we could identify who the polluters are, the principle would not (yet) address all historical pollution and there are serious questions about whether the polluters' payments can compensate for the damage to the climate their pollution causes.

How much should polluters pay?

Let us now turn to the issue about the amount polluters should pay. We are witnessing the associated dangers of climate change today. This gives us reason to act. However, current climate change is a result of earlier practices by past generations. How much we should pay must address this historical dimension.

The British economist and House of Lords Peer, Nicholas Stern, argues that greenhouse gas emissions are an example of "the greatest market failure the world has ever seen."[18] He is correct. This is because the prices of goods, such as petrol, do not reflect the true costs to society of their production and use. The full use of petrol is more than our filling a tank, but the carbon emissions it will produce and their remaining in the atmosphere for decades afterwards. The polluter pays principle typically focuses on consumption, but not production which is a mistake: we should focus on both.[19] These full costs include the damaging effects of our emissions on the climate.

Polluters who create carbon emissions through their oil consumption share responsibility for the full cost of pollution with the oil refineries that produce oil for the market, for example. So if we argue that creating carbon emissions may entail having to pay for the pollution, then we should recognize that carbon emissions arise with production and consumption. Both must be reflected in calculating how much we should pay.

Characterizing the problem in this way highlights a ready solution: ensure the full costs of producing and consuming oil products is reflected in their price. A commonly shared view among normative political theorists is that oil producers should be left to set their own prices subject to taxation by government. One reason is that oil producers endure the costs of production, not government. Therefore, government should not dictate the exact costs of producing barrels of oil.

However, oil producers are not as well placed as government to address the consumption of oil products. This is because of the nature of carbon emissions. Governments are more able to administer mitigation policies and attempt to coordinate them with other states through international regulatory agreements. This is all the more important because taxation might have certain costs for companies today, but the costs of inaction are spread out globally and internationally raising costs for us all tomorrow. Someone must pick up the bill and so it is widely argued by normative political theorists that some form of tax should be imposed on the purchase of oil products.

While Stern primarily focuses on carbon – and the promotion of a "carbon pricing" as most human-created climate change is from carbon emissions – the polluter pays principle can expand beyond payments for oil consumption alone.[20] The main issue is not whether pollution is carbon-based *per se*, but rather whether the pollution contributes to some environmental harm such as climate change and its potentially harmful effects. Emissions of other products, such as methane, may also be relevant for a polluter pays principle to consider. So Stern's general position remains applicable although it is focused only on carbon emissions.

This is an important point because not all emissions are the same. Our emissions today have greater impact because of past emissions already polluting our current atmosphere.[21] This means that the true costs of carbon emissions is higher – and more difficult to calculate.

The polluter pays principle attempts to capture the costs of carbon and greenhouse gas emissions so it may provide compensation. There are several serious difficulties with this project. The first problem is that the amount of the full costs of production and consumption will be difficult to calculate in a way that ensures that compensation is accounted for properly. The costs of consumption are intended to include the effects of my emissions. Yet, these effects may linger for a long period perhaps long after my death. As further emissions build off of earlier emissions, the amount of damage present and future emissions contributes to will increase. This is a complex picture of rising costs that polluters must pay at an increasing rate.

A second, more serious problem is the pricing is supposed to incentivize conservation through reduced emissions. The higher costs of polluting should lead to less polluting activities. So the costs need to be set at a price that will have this effect. This price effectively acts like a deterrent and looks to future behavior. Contrast this way of setting costs by instead looking only at what is required to compensate for the damage done. This different way of calculating the price polluters should pay acts like corrective justice looking at what has happened in the recent past. The polluter pays principle aims to both correct for damage caused while deterring polluters to emit below a sustainable level globally. These are different considerations looking in opposite directions. It is unclear they each lead to setting the exact same price – and so the polluter pays principle has a conflict of principles at work in how it sets the amount that polluters should pay.

There is a problem of distributive justice, too. Poorer countries use less oil so would be charged less while more affluent countries use greater quantities.[22] But what might be the effect of setting a higher price? Those wealthy countries best able to afford it could manage, but poorer countries – already relatively less reliant – would be impacted much more at a higher price. The same costs for rich and poor will have very different impacts for each that look like they would embed existing global distributive injustices making the poorer countries worse off. Yet, if we then charged a lower price for poorer countries to reduce the economic impact on them, we would then set a price lower than what is required for compensating for the harms produced by carbon emissions. While reduced emissions are surely a necessary part of any serious climate change strategy that is supported by making the use of emissions more expensive, the polluter pays principle does not come close to living up to its own name.

And then there is the sometimes wildly fluctuating price of oil depending on changes in the market. For defenders of the polluter pays principle, we can tax our way to climate change justice and a sustainable future. While higher petrol costs are associated with fewer emissions, there is no strong evidence to suggest that these emissions would drop sufficiently low to yield the kind of sustainable, conservationist future the polluter pays principle promises for only a few dollars extra per barrel as some proponents endorse.[23] The kind of outcomes they support may require more significant taxes than they claim.

Furthermore, the lockdowns across multiple countries to combat the global coronavirus COVID-19 outbreak this spring saw hundreds of millions of people avoiding any unnecessary travel. Many worked or studied from home if they could. This saw the taps on oil wells turned off as storage facilities were full with few people buying as they stayed at home during lockdowns. People's behaviors changed, but not because of the price. In April 2020, oil lost so much value that "traders were being paid more than $40 to buy a barrel of oil" as stocks were dumped after demand had slumped.[24] This is an example where use does not closely link to price especially during emergencies – and we are in a climate emergency now.

Is the polluter pays principle anti-conservation?

A major appeal in policy circles is the polluter pays principle's aim of enabling sustainable greenhouse gas emission levels. However, the principle cannot guarantee this. It is often proposed as a tax on oil consumption. The idea is a tax would raise the costs of using oil and so lead to sustainable consumption of fossil fuels. While higher costs are associated with fewer emissions, there is no evidence to suggest that they will drop sufficiently low.[25]

Moreover, *the polluter can pollute as much as he or she can pay for*. If polluters are assumed able to pay for the environmental costs from their emissions, then global carbon emissions may rise where polluters have the ability to pay since there is no cap on how much emissions may be consumed. For these reasons, the polluter pays principle may claim to be a mitigation strategy but it is not satisfactorily conservationist. Global emissions can rise far above any sustainable level.

Do emission caps remove negative duties?

Some polluter pays principle advocates are sensitive to these objections. For example, political theorist Simon Caney argues that we are all under a duty not to exceed our quota for greenhouse gas emissions.[26] Global emissions must be capped. The idea is that the cap should be set so low as to secure a sustainable future while revenue generated from the tax on emissions can support mitigation efforts. In this way, Caney attempts to support the polluter pays principle within a cap on global emissions.

One problem is that the principle loses its motivational force. A negative duty to compensate for potential risk of environmental harm may be more compelling than a positive duty to compensate despite the absence of risk. If global emissions were capped to ensure conservation, then polluters would not need to compensate others for any harm because none would arise within the global emissions cap. The polluter pays principle is reduced to a positive duty with the aim of generating resources through fundraising to assist conservation – and not as a negative duty aim to compensate for harm.

To restate this problem: the principle is founded on the idea that polluters harm others by polluting and so they should pay. But if emissions are capped so that no polluter harms others by polluting, then it becomes unclear why polluters have a duty to compensate others because they would not have a negative duty to do so: no compensation would be required because no harm would be done.

There is a further concern about such a "cap-and-trade" system. The benefit is it would set an absolute ceiling for carbon emissions. But as Eric Posner and David Weisbach point out, this might "create emissions certainty but would not create environmental certainty."[27] Climate sensitivity is not as fixed as pinpointing emission concentration levels. In other words, we might be able to control the emissions produced but cannot guarantee that, once emitted, the environment will respond in any specific way. The certainty of emission levels is more certain than that of the emission's effects.

Should we pay polluters to stop emissions?

There is an alternative, inverted principle some like past and present University of Chicago Law School colleagues Eric Posner, Cass Sunstein and David Weisbach who support a "polluted be paid principle," or which we might call *the pay polluters to not pollute principle*.[28] This is the idea that major greenhouse gas emitters, specifically like the United States, should be paid to sign up to any treaty. This payment would be made in return for the United States reducing its emissions – and paid primarily by developing countries. The argument is that developing countries are more at risk from continuing emissions than countries like the United States. It is in the interest of developing countries to agree such "side-payments" to ensure a treaty agreement on climate change is also firmly in the interest of the United States.

This principle is partly motivated by rejecting the idea that any use of carbon or fossil fuels requires compensation. This is on the grounds that "emitting carbon is part of everyday life."[29] In their view, it is no sin to emit – only to emit too much.[30]

But the principle is primarily motivated as ensuring *feasibility*. They claim this is about aligning different interests in support of the same end – a comprehensive climate treaty. It is in the interests of developing countries for affluent states to reduce their emissions because developing countries are more vulnerable to climate change effects. However, it is thought that reducing emissions and curtailing climate change is not yet in the short- or medium-interests of the United States. Further climate change will impact on coastal communities, but also make more land agrigable, such as in currently not-so-agrigrable places like Alaska. Thus, weighing up the costs and benefits in a cost-benefit analysis might help explain why the United States has taken a back seat to other countries and regions like the United Kingdom or European Union on climate issues. Warming will harm some parts of the U.S., but free up new currently nonusable land for agriculture. At a macro-level, climate change might provide short- to medium-term benefits overall. Or so Posner, Sunstein and Weisbach argue.

In recommending its feasibility, there seems to be a greater emphasis on feasibility from the perspective of the United States over the views of others. For example, it is easily conceivable that poorer developing countries more vulnerable to climate change's effects would strongly resist having to agree to payments to far wealthier countries so they would stop damaging the global climate. Stephen Gardiner likens this situation to extortion. Ultimately, this is about paying others not to pollute and cause significant harms globally. Not about compensating against future gains foregone, but extortion where victims pay the wrongdoer to hold off threatening their very survival.[31] For this reason, there are serious ethical problems with a principle based on paying off the polluters rather than requiring them to pay for the problems their pollution creates.

To be fair, Weisbach and his colleagues favor the United States reducing its emissions to zero in recognition of the urgent, significant action that must be taken.[32] Their concern is that a global treaty is needed and a feasible pathway must be found so that all countries are signed up without others noncooperating.[33] The reason is not because ethics is thought to provide a useful guide to policy making, but a matter of tracking self-interest. Weisbach says: "if people or nations only pursue their self-interest, nothing ethics can say will influence the outcome."[34] Of course, the study of ethics can help us sharpen what is our "self-interest" and the value we place in short-, medium- and longer- term timescales. For example, it is not universally accepted that self-interest means atomistic individualism.

And even if this were untrue, it is surely an *ethical* matter how much weight is to be given to my *immediate* satisfaction of self-interested goals versus *longer-term* satisfaction as this is a matter of values. It is no given that immediate gratification and short-term profits are the only ends of self-interested pursuits. If my enjoying increased wealth at the expense of others today means we race more quickly to an environmental catastrophe for me or my family and friends where none can enjoy this wealth, it is difficult to see why self-interest cannot take any account of its social context and timescales – unless, of course, the specific notion of "self-interest" being used is itself an ethical position dressed up as fact while grounded only on fact-free presuppositions or an arbitrary definition. For the above reasons, the pay polluters to not pollute principle is unconvincing as a politically or morally feasible option.

Global Resources Dividend

A popular version of the polluter pays principle among political philosophers is Thomas Pogge's *Global Resources Divided*.[35] Much of the background argument will be familiar as it echoes the common justifications used for the polluter pays principle more broadly. He argues that the effect of oil consumption is generally that of affluent states consuming oil products and producing carbon emissions that present less affluent states with more significant threats to less affluent states: this is because less affluent states are more likely to have populations in severe poverty. Affluent states receive benefits of improving economies at the expense of less affluent states left to bear more of the costs and more exposed to the threats that climate change creates.

Pogge argues that the citizens of affluent states have a negative duty to provide some effective means of compensating persons in less affluent states. This is because "we" in affluent states knowingly, foreseeable and avoidably engage in practices that lead to "our" benefitting at the expense of others who are harmed by their bearing more of the harms and risks of climate change among other problems.[36]

He claims there should be a Global Resources Dividend (GRD). This is essentially a tax of perhaps $2 per barrel of oil that is collected by government and used to provide compensation to less affluent states by helping provide mitigation so they are protected from the potential harms of climate change. Pogge argues the amounts generated through this dividend will more than compensate less affluent states and also contribute to fewer carbon emissions by raising the cost of petrol.[37]

Pogge says:

> The GRD proposal is meant to show that there are feasible alternative ways of organizing our global economic order, that the choice among these alternatives makes a substantial difference to how much severe poverty there is worldwide, and that there ae weighty moral reasons to make this choice so as to minimize such poverty.[38]

This comment is revealing. The proposal's main purpose is actually as a means of global distributive justice providing new funding for severe poverty alleviation in developing countries. Pogge is part of a growing number of commentators that see the opportunity to address the challenges of climate change as a time for helping tackle severe poverty. He is not alone in making this link.[39] Even if the pursuit of reducing emissions is different from alleviating severe poverty, there is something attractive in trying to make progress against two major global challenges at once.

Nonetheless, Pogge's GRD is too weak to serve as a polluter pays principle. While it is motivated, in part, by the same view of seeking to compensate those impacted from the harms of greenhouse gas emissions, the tax put on oil seems set far too low to either sufficiently drive down demand to below a sustainable level nor does it appear able to raise enough funds to cover the full costs of carbon emissions. Moreover, its end is not to spend the funds raised on climate mitigation or adaptation strategies but on alleviating severe poverty. For many, this will seem like the right choice. But the problem is – as a mitigation strategy – it fails to provide the sustainable option that it promises.

Border tax adjustment

Rolf Weber proposes a different, but not entirely unrelated taxation regime to secure greater climate change justice: the border tax adjustment (BTA).[40] This is a means to compensate for lax environmental standards in international trade. He argues it can be realized through two different models. The first is through the levy of a tax on products from states with "lower production standards" or the refund of domestic taxes where these products are exported. The second is "the requirement to obtain emission allowances and to participate in an Emission Trading System."[41]

The BTA measures are set by national governments and assessed by the World Trade Organization (WTO). Weber argues that BTA measures should be designed "as indirect product taxes within the scope of Article II GATT" where these "taxes aim at creating a level plain field between 'like products' in the country of destination."[42] He adds: "if the carbon tax can not be linked to the goods directly, but only to the production of the goods, it is important to design the tax in a way that it has an internal or incorporated nature, similar to charges on fuels or energy inputs on fossil fuels used in the production process."[43]

One key issue concerns noncompliance. If BTA is going to work, then it must be implemented by national governments while not running afoul of WTO regulations. But what if other governments did not cooperate with the ETS? What body would be empowered sufficiently to address issues of noncompliance and would it be effective? Of course, some cases of a lack of consensus may be comparatively more tractable than others: disagreements about arbitrary discrimination under trade law might be easier to resolve than disagreements concerning climate policy. Nonetheless, the answers to these questions are unclear.

A second issue is scope. The plan focuses on international agreements between states. These are designed and implemented by national governments in relation to trade across state boundaries aiming to address the problem of carbon leakages across borders. However, trading patterns can be more complex.

For example, a customer in the United Kingdom could buy services advertised on a website hosted in the United States, but where any cash transactions are made in the Republic of Ireland or Luxembourg to be transported using a Dutch delivery service. While this illustration is imaginary, what would look like someone purchasing a product from a website and having it delivered to their door can involve multiple points at which different states have relevance. Some relate to web design and hosting, others the registered office and still others the distribution arm. How might BTA legislated by a single country relate to the complex web of market relations that governs an increasing part of the contemporary globalization we find ourselves? This raises further questions about our ability to police noncompliance in an interconnected world of firms most of whom are small, but many of whom have larger market shares than many other countries.

One possible response is to say a BTA might be imperfect, but it is better than our doing nothing. It might be argued that some agreement is preferable to none. This is because the use of the BTA could help win wider support over time leading to a more stringent international agreement.[44] The BTA might be understood as a means to an important end of delivering a climate policy more fit for purpose. So perhaps the scope is limited, but the BTA makes possible a wider inclusion of countries in future. This is guesswork.

However, it is reasonable to believe that if the BTA can be made workable at the national and regional level then it might achieve greater agreement. International agreements carrying universal support are very difficult to achieve from the start and normally require a slower build-up of support and momentum over time. The problem of scope is a concern, but need not be a deal-breaker if we accept BTAs as a means to a better end. But there remains the possibility that a dedicated focus on BTAs to the exclusion of a more inclusive international deal could incentivize states to pursue unilateral agreements instead of multilateral, comprehensive agreements for collective action on climate policy.[45]

A final issue is effectiveness. What likely effect can we expect from a BTA? If we accepted its use and resolved issues concerning compliance and noncompliance, what should it look like in practice to deliver the practical benefits it aspires to achieve? This is highly relevant to our agreeing to BTA as at most a second-best solution or, indeed, a solution worth pursuing at all.

It is unclear why we should ultimately support BTAs, assuming their potential legality, given the many "known unknowns" (to use a phrase by Donald Rumsfeld) concerning its use.[46] So perhaps implementing a BTA is possible, but this is not to establish it is preferable in relation to other options. This is especially true in light of a lack of international coordination on what the terms of the BTA should be in order to (second-)best mitigate climate change effects.

One response might be to reject the BTA in favor of some form of tax on carbon emissions that applies universally.[47] Such a tax would create a more level playing field by including a similar standard for all. This could be a means of raising funds that could help finance a transition to a low-carbon economy. It would become more profitable for firms to reduce their carbon footprint in their production and sale of goods and services.

While this approach might address concerns about scope, it is not obvious that it would be more effective than alternatives such as the BTA. For example, the fact that costs might rise when producing certain goods for all will not impact everyone the same. Those who can afford goods and services made more expensive may still do so. The problem is akin to my criticism previously of the polluter pays principle more generally in that it might permit polluters to pollute as much as they can pay. Treating all states the same might sound like sharing responsibilities for tackling climate change equally. But higher taxes may not yield desired outcomes on their own.

Furthermore, not all states possess the same historic responsibilities for the problems we face due to climate change. So treating all states the same denies the greater responsibilities some have for the problems faced.[48] Finally, the use of a tax like the BTA does not necessary shift the burden of responsibility for climate change to emissions producers because they might pass the extra costs to consumers.[49] For these reasons, it cannot provide the solution to our climate change challenges that it claims.

Conclusion

The polluter pays principle is a proposal highly popular with policy-makers and theorists alike for how the challenges of climate change can be met. There is no doubt that as a partial measure this principle can play a crucial role in helping disincentivize the use of fossil fuels through increased costs and raising valuable new sources of funding for mitigation – and adaptation – efforts. In the short term, having a polluter pays principle is better than not having it.

However, the principle is an unsatisfactory solution to how we might best address the associated dangers of climate change – and it fails to provide any guarantee of longer-term sustainability. While the principle appears simple, its implementation is complex and it has difficulty in identifying who the polluters are that should pay, how much should be paid, who should be paid and for what and, finally, how this would secure lower emissions when polluters may emit as much as they are able to pay. To find a longer-term solution, we will have to explore further options.

Notes

1 See Simon Caney, "Cosmopolitan Justice, Responsibility, and Global Climate Change," *Leiden Journal of International Law* 18 (2005): 747–775; S. Gaines,

"The Polluter-Pays Principle: From Economic Equity to Environmental Ethos," *Texas International Law Journal* 26 (1991): 463–495; Giddens, *The Politics of Climate Change*, 92; Eric Neumayer, "In Defence of Historical Accountability for Greenhouse Gas Emissions," *Ecological Economics* 33 (2000): 185–192; and Henry Shue, "Global Environment and International Inequality," *International Affairs* 75 (1999): 533–537.

2 Not all conservationists endorse the ecological footprint, fair shares or polluter pays principle approaches. My focus on these proposals is because the overwhelming majority do support them as their leading contributions to debates over how to act and why over climate change.

3 See Paul Baer, "Adaptation: Who Pays Whom," in Neil Adger et al. (eds.), *Fairness in Adaptation to Climate Change* (Cambridge: MIT Press, 2006): 131–154.

4 See Caney, "Cosmopolitan Justice, Responsibility, and Global Climate Change," 755. On the problem with states and responsibility, see Onora O'Neill, "Agents of Justice," *Metaphilosophy* 32 (2001): 180–195.

5 This brings to mind Numbers 14:18 where it is said the sins of the father will be visited on their children. (See also Job 21:19 which takes a different view for those interested in this perspective.)

6 James Garvey, *The Ethics of Climate Change: Right and Wrong in a Warming World* (London: Continuum, 2008): 115.

7 See John E. Tilton, "Global Climate Policy and the Polluter Pays Principle: A Different Perspective," *Resources Policy* 50 (2016): 117–118.

8 See Kyoto Protocol to the UN Framework on Climate Change, Kyoto, Japan, December 10, 1997, in force February 16, 2005, 2303 UNTS 148.

9 See Thom Brooks, "Cosmopolitanism and Distributing Responsibilities," *Critical Review of International Social and Political Philosophy* 5 (2002): 92–97 and Thom Brooks, "Collective Responsibility for Severe Poverty," *Global Policy* (forthcoming).

10 See J. Spencer Atkins, "Have You Benefitted from Carbon Emissions? You May Be a 'Morally Objectionable Free Rider'," *Environmental Ethics* 40 (2018): 283–296.

11 See Gardiner, *A Perfect Moral Storm*, 419.

12 See Michael Sandel, "Should We Buy the Right to Pollute?" in *Public Philosophy: Essays on Morality in Politics* (Cambridge: Harvard University Press, 2005): 93–96.

13 See Daniel Yuichi Kono, "Compensating for the Climate: Unemployment Insurance and Climate Change Votes," *Political Studies* 68 (2020): 167–186.

14 My thanks to Melissa Lane for highlighting this problem.

15 See Eric A. Posner and David Weisbach, *Climate Change Justice* (Princeton: Princeton University Press, 2010): 81.

16 See Posner and Weisbach, *Climate Change Justice*, 47.

17 One compelling approach to thinking about harm, future generations and the non-identity problem is offered by Joseph Mazor. He argues that present persons have justice-based obligations to each other to conserve natural resources for future generations where these generations as understood as "a chain of overlapping generations" rather than monoliths. See Joseph Mazor, "Liberal Justice, Future People, and Natural Resource Conservation," *Philosophy and Public Affairs* 38 (2010): 380–408 and Edward A. Page, *Climate Change, Justice, and Future Generations* (Cheltenham: Edward Elgar, 2006). On the nonidentity problem

more generally, see Derek Parfit, *Reasons and Persons* (Oxford: Oxford University Press, 1984): 351–380.

18 Stern, *A Blueprint for a Safer Planet*, 11.

19 See Stern, *A Blueprint for a Safer Planet*, 159.

20 See Grantham Research Institute on Climate Change and the Environment, "What Is a Carbon Price and Why Do We Need One?" (May 17, 2018), url: www.lse.ac.uk/GranthamInstitute/faqs/what-is-a-carbon-price-and-why-do-we-need-one/.

21 See Gardiner, *A Perfect Moral Storm*, 39, 419.

22 See Yao Li, Jin Fan, Dingtao Zhao, Yanrui Wu and Jun Li, "Tiered Gasoline Pricing: A Personal Carbon Trading Perspective," *Energy Policy* 89 (2016): 194–201.

23 U.S. Energy Information Administration, *International Energy Outlook 2011* (Washington, DC: Energy Information Administration, 2011): 6–7, url: www.eia.gov/forecasts/ieo/emissions.cfm/.

24 See Sky News, "Coronavirus: US Oil Price Plunges Below Zero for First Time in History as Outbreak Hits Demand" (April 20, 2020), url: https://news.sky.com/story/us-oil-price-plunges-below-zero-for-first-time-in-history-11976165.

25 See U.S. Energy Information Administration, *International Energy Outlook 2011*, 6–7.

26 See Caney, "Cosmopolitan Justice, Responsibility, and Global Climate Change," 769. While I am not convinced by the argument, Caney has developed, to my mind, the most sophisticated and compelling of the very many polluter pays principles in circulation. He is also an old friend to whom I will always be grateful for changing positions which afforded me my first full-time academic job. I wore as a badge of pride that I was his replacement in post. No doubt we are cut from somewhat different cloths. Our former university in Newcastle upon Tyne somehow makes do without either of us now.

27 See Posner and Weisbach, *Climate Change Justice*, 53.

28 For example, see Posner and Weisbach, *Climate Change Justice*, 84–88 and Eric A. Posner and Cass R. Sunstein, "Climate Change Justice," *Georgetown Law Journal* 96 (2008): at 1569.

29 Posner and Weisbach, *Climate Change Justice*, 55.

30 This point echoes the distinction of necessary (or "subsistence") and unnecessary (or "luxury") emissions made by Henry Shue. There is no issue with necessary, or subsistence, emissions. In fact, he claims there is actually a right to them. The only issue is with emissions above that necessary level. In a similar vein of argument, Posner and Weisbach appear to claim that not all emissions are necessary problematic *per se*.

31 See Stephen M. Gardiner, "Justice vs. Extortion," in Stephen M. Gardiner and David A. Weisbach (eds.), *Debating Climate Ethics* (Oxford: Oxford University Press, 2016): 87–133.

32 See Weisbach, "Climate Policy and Self-Interest," 170–200.

33 See Weisbach, "Climate Policy and Self-Interest," 195.

34 Weisbach, "Introduction to Part II," 152.

35 See Thomas Pogge, *World Poverty and Human Rights*, 2nd ed. (Cambridge: Polity, 2008): 202–221.

36 Thom Brooks, "Punishing States That Cause Global Poverty," *William Mitchell Law Review* 33 (2007): 519–532.

37 See Pogge, *World Poverty and Human Rights*, 202–221.

38 Pogge, *World Poverty and Human Rights*, 197.

39 See also Stern, *A Blueprint for a Safer Planet*, 8.
40 This section builds off of the arguments in Thom Brooks, "Climate Change Justice through Taxation?" *Climatic Change* 133 (2015): 419–426.
41 See Rolf H. Weber, "Border Tax Adjustments–Legal Perspective," *Climatic Change* 133 (2015): 407–417.
42 Weber, "Border Tax Adjustment."
43 Weber, "Border Tax Adjustment."
44 Dominic Roser and Luke Tomlinson, "Trade Policies and Climate Change: Border Carbon Adjustments as a Tool for a Just Global Climate Regime," *Ancilla Iuris* (2014): 222–244, at p. 229.
45 Roser and Tomlinson, "Trade Policies and Climate Change," 242.
46 As much as I continue to dislike his political judgement and press briefings, I am to my shame continually referring to "known knowns," "known unknowns," "unknown knowns" and "unknown unknowns" ever since they were first used.
47 See Robyn Eckersley, "The Politics of Carbon Leakage and the Fairness of Border Measures," *Ethics and International Affairs* 24 (2010): 367–393.
48 See Clara Brandi, "Trade and Climate Change: Environmental, Economic and Ethical Perspectives on Border Carbon Adjustments," *Ethics, Policy and Environment* 16 (2013): 79–93.
49 Roser and Tomlinson, "Trade Policies and Climate Change," 236–237.

3 Adaptation

"What is the use of a house, if you haven't got a tolerable planet to put it on?"

– Henry David Thoreau

Reducing emissions isn't enough

The previous chapters explored different proposals for how we might reduce greenhouse gas emissions. There is no doubt that these emissions contribute to climate change, and there is a strong, global consensus they must be cut significantly. However important making these big cuts are to facing the challenges posed from climate change, these reductions are not adequate on their own – and, as we have seen, mitigation proposals ranging from the ecological footprint to the polluter pays principle all fall short of where we need to be.

We need *to adapt* to a world that is already changing.[1] It is crucial to note upfront that it is widely, and correctly, understood that adaptation is *a reality* and *not optional*.[2] For example, the planet's average temperature is expected to rise by at least 1.5°C this century – and quite possibly to 2°C or higher. The difference between a 1.5°C or 2°C will determine whether 70 percent of coral reefs will die or 99 percent, whether the Arctic Ocean is bare of sea ice once every century or each decade and much more.[3] Big changes are forecast in our nearing future and we will have to adapt to our changing climate.

Most pro-mitigation conservationists accept that any compelling climate change policy must incorporate adaptation because any mitigation alone "will not be enough."[4] Others accept adaptation is needed to help us in is "making the best of a bad job."[5] Or as a Royal Society report once put it "the lack of progress of the political processes" – and not the search for the "safest and most predictable" approach of cutting greenhouse gas emissions – motivates adaptation.[6] Whether through necessity or pragmatism, adaptation offers proposals that must be taken seriously.

Stephen Gardiner correctly argues:

> The first thing to note . . . is that adaptation measures will clearly need to be part of any sensible climate policy, because we are already committed to some warming due to past emissions, and almost all of the proposed abatement strategies envisage that overall global emissions will continue to rise for at least the next few decades, committing us to even more.[7]

Adaptation may play an important role in formulating climate change policy alongside real reduction in carbon emissions. The question is whether adaptation is an appropriate primary solution to the problem of climate change instead of mitigation – and if it can provide us with the "solution" to climate change its proponents champion. This chapter explores the main ideas, key proposals and assesses the prospects for adapting ourselves to a sustainable future.

Varieties of adaptation

Adaptation measures may take several forms, including:

- Carbon dioxide removal, reforestation and direct capture of carbon from the air
- Greater energy efficiency so our electrical appliances and devices require fewer emissions to work
- Improved heating insulation at home
- Investment in public transport to reduce reliance on using private transport
- Increasing recycling and promoting refusing
- Ecosystem-based adaptations, such as planting mangroves to provide natural flood defenses, well-protected lakes to retain water sources in times of drought and healthy forests to reduce risks of wildfires
- Walking or cycling instead of using a car
- The use of labelling and nudges to help shift common behaviors promoting a more climate-friendly lifestyle

This list is not exhaustive, but homes and transport loom large as key areas where emissions are created by us all. It shows the wide range of activities that make up adaptation to climate changes. Many can become easily incorporated into our everyday lives and they are not always the products of science fiction that we can sometimes find, such as ideas for floating islands to create new housing so we might adapt to rising sea levels and the like.

Adaptation is about changing our relationship to nature as our climate changes. So, for example, the potential threats to coastal communities from rising sea water may be addressed through building better flood defenses or relocation. Climate change will see land become more arid and less agriculturally productive. We may address this problem through the greater use of genetically modified crops that may better thrive than traditional crops. Tropical diseases spreading to new geographical areas might be approached through inoculation measures. These measures can provide us with a sustainable future without engaging in costly major emissions reductions. While every part of our planet will be affected, it will not be uniform for every place and all individuals.

Adaptation can make a significant difference, too. An example is the Great Green Wall project planting a 4,815-mile barrier of trees like drought-resistant acacia, baobab and Moringa with the aim of halting the spread of the Sahara.[8] This $8 billion reforestation of 247 million acres from Dakar to Djibouti will absorb about 250 million metric tons of carbon dioxide when it reaches its scheduled completion in 2030. The fear is that the desert sands will spread without projects like this, with an estimated 85 million in sub-Saharan Africa forced to migrate.

One controversial form of adaptation is *geoengineering*, where attempts are made at re-engineering on a truly global scale. There is more than half a century of research into how this might work. Perhaps the most well-known illustration of these efforts at what is called "solar radiation management" through sulfate injection.[9] This works by loading airplanes with sulfate particles and having them spray their loads at 65,000 feet in several thousand flights each year. The effect of the particles is to deflect solar radiation and create a cooling effect mimicking the output of volcanic eruptions.

But there are other effects. Sulfate injection reengineers a part of the atmosphere to produce a small, but important, degree of cooling at a potentially stratospheric price. Putting particles in the air – which will only remain in the clouds for a couple years and requiring annual top-ups – does not touch the continuing accumulation of greenhouse gas emissions. While the planet might slightly cool, damaging effects like ocean acidification and potential major unknown shifts in weather patterns could easily create more problems than this effort attempts to solve.[10] Plus, such a global change would seemingly require a global consensus that seems very elusive.[11]

When we think about adapting to a warmer, less hospitable planet the first things to come to mind are melting ice caps, rising sea levels and the prevalence of droughts. Many of our discussions and policy proposals focus on relocation, flood defenses and the development of genetically-modified crops. Too often missing from view is the fundamental issue of sanitation, including clean water, sewage treatment and waste disposal that are essential to any long-term adaptation strategy. We do not only inhabit homes and eat, but we require safe drinking water and climate-friendly waste management.[12]

The World Health Organization estimates a quarter of a million additional deaths due to climate change each year between 2030 and 2050 with many linked to preventable diseases associated with inadequate sanitation and hygiene.[13] More than a quarter of all water use in the home is literally flushed down the toilet.[14] Ensuring sustainable sanitation matters. Finding ways of reducing such usage and its impact on the climate, such as using collected rainwater for use in bathrooms, can make a significant difference.

The French poet and dramatist Victor Hugo remarked in 1862 that "all the human and animal manure which the world wastes, if returned to the land, instead of being thrown into the sea, would suffice to nourish the world."[15] This vision is coming true as more work is being done in transforming our view of human urine and manure from waste to be disposed to a resource that can be used.[16]

Adaptation measures are typically understood anthropocentrically: adaptation puts people first.[17] So we speak of relocating *human* communities or genetically modifying food to feed human beings. There is little consideration of how the natural world might be better adapted to ensure continued flourishing. Adaptation is primarily about how "we" might best adapt to climate change. There is less concern about how plant and animal species may be affected by climate change than found with conservationist approaches in general.[18]

Perhaps the best exemplar – and most thought-provoking – example is work by philosopher and bioethicist S. Matthew Liao. Nearly a decade ago, he coauthored a landmark paper that proposed a new way for dealing with the ever- increasing impact and consumptions of a growing human population.[19] They proposed *human engineering*: if we can change our underlying biology by altering our size or diet, then we could create greener humans with a reduced environmental impact.

For example, we know that raising livestock for human consumption amounts to about 18 percent of greenhouse emissions. If we are too weak-willed to avoid the desire to eat beef, it could be made possible should we engineer people to dislike the taste of hamburgers. And if we reduced our physical size by 15 centimeters, it would lead to 25 percent less mass – less to be fed or watered. While Liao proposes this more as a hypothetical future and not a call for mass human engineering along these lines, his ideas open up new options for how we might adapt to climate changed future.

To summarize, adaptation is commonly focused on the ways and means we might adapt ourselves to a changing climate. There a wide array of forms that adaptation can take from walking or recycling to direct carbon capture and geoengineering. While we do not want for options, is a sustainable ever- after possible through adaptation? We turn to this issue next.

The audacity of adaptation

Adaptation proponents believe it can deliver a win-win scenario. It is easy to be struck bordering on surprise at the overconfidence of how well we can adapt successfully.

One limitation is future uncertainty about the environment.[20] What future must we adapt to? We must especially have a clearer conception about what a future of adaptation might look like in contrast to its mitigation alternative. While there are models of likely effects from climate change in our near future, these models become far more speculative the further ahead we look. Adaptation is a strategy for a future world about which we lack sufficient clarity and certainty.

A second limitation is future uncertainty about the likely success of adaptation measures. Suppose we could have confidence in models of future environmental conditions should we choose adaptation over mitigation. The problem is that we cannot safely test proposals in the way that many biologists or chemists might conduct experiments in a controlled laboratory. Any measures would have some real degree of uncertainty of success beyond an acceptable level given the very high costs of failure. Indeed, many proposed adaptation measures have either not been tried or do not yet exist.[21] They peer into the future and can only speculate.

For example, some argue in favor of "carbon capture" where carbon is removed from the atmosphere and pumped into depleted oil fields deep underwater.[22] Carbon is also captured and used in manufacturing products like baking soda and construction materials. In this myriad of ways, adaptation efforts have been made to reduce existing carbon and store it safely.

The problem – especially with deep underwater storage – is that the future risk to human and marine life is unknown with potentially deadly consequences for both.[23] Some argue it is unfeasible to avoid the planet warming by 2°C or less by 2100 without carbon capture storage.[24] But yet it seems impossible to guarantee any possibility of future leakage and difficult to assess impact of damage from leaks – and hard to predict the costs; agents might not consider future costs as any leakage could transpire many decades from now further complicating the potential costs financially or otherwise over time.

Many have put great faith in technology to help address climate change. For example, Matthew Kahn claims: "in a world with billions of educated, ambitious individuals, the best adaptations and innovations will be pretty good."[25] Adaptation's popularity lies in its being a "cheap and simple" solution.[26] However, Dale Jamieson says:

> Technological approaches are popular both with politicians and with the public because they promise solutions to environmental problems without forcing us to change our values, ways of life, or economic systems . . . the image of the scientist as the "can-do" guy who can solve any problem remains quite potent.[27]

Cheap solutions that will leave our daily lives largely unchanged may always be an attractive option as the easiest choice requiring little effort.[28]

The problem is that too often technological advances producing energy savings have been counterproductive – and so their future effectiveness is somewhat unreliable.[29] For example, it is argued:

> More power-efficient washing machines or better insulated houses will help the environment; but they also cut our bills, and that immediately means we lose some of the environmental gain by spending the saved money on something else. As cars have become more fuel-efficient we have chosen to drive further. As houses have become better insulated we have raised standards of heating, and as we put in energy-saving light bulbs the chances are that we start to think it doesn't matter so much leaving them on.[30]

While great advances have been made in energy efficiencies, these have not led to the carbon emission reductions they promised. If adaptation were to be a primary aim of climate change policy, then we require higher confidence that technological advances would lead to greater reductions rather than result in counterproductive behavior.

These new technologies are often not cheap either. Stephen Gardiner observes that green energy comes with a higher price tag than its carbon-fueled cousins – and "green energy is a luxury that the world's poor will not be able to afford."[31] So even if such technologies could deliver on their bold promises and better adapt us to a changing climate (and we have reason to doubt this), the direct benefits would be most felt by those least vulnerable to climate change's damaging effects, namely, affluent countries in the Global North. Developing countries would be more likely to miss out.

Adapting trade-offs

Adaptation advocates share several views in common. The first is less certainty that major reductions in carbon emissions are strictly necessary: "we will save ourselves by adapting to our ever-changing circumstances. . . . At the end of the day, the story will have a happy ending."[32] While there is no doubt that climate change is taking place, there is a greater certainty that we can effectively adapt to the changing conditions that we will face in future and in sufficient time.[33] Major adaptation efforts, such as greater urbanization and reliance on genetically modified foods and nuclear energy, can provide a satisfactory solution.[34]

One reason why they argue that we should focus more on adaptation rather than conservation is because the former is a more cost-effective approach. For example, Bjorn Lomborg says: "it will be far more expensive to cut CO_2 emissions radically than to pay the costs of adaptation to the increased temperatures."[35] Mitigation approaches are estimated to cost about 2 percent of GDP per annum (or roughly $1 trillion per annum).[36] Our resources are better spent adapting ourselves to climate change and spending our savings on other major social issues, such as poverty alleviation.[37] We should reduce emissions to a level of sustainable adaptability. This will cost less than reducing emissions levels to a sustainable level not requiring adaptation measures. The savings from this might be used to do more good for the global poor than mere protection from the associated dangers of climate change.[38]

In a similar vein, Eric Posner and David Weisbach claim that tackling climate change has important trade-offs. For example, they note that malaria is a disease of the poor. They cite the World Health Organization that people in countries with annual incomes of $3,000 or more do not die of malaria. In climate models used by the IPCC, they predict all regions of the world will surpass the $3,000 threshold by 2085. Yet, if developing countries diverted resources to climate change reduction, incomes will be systematically lower, delaying the lifting up of all countries about the $3,000 mark and exposing more people to a risk of death from malaria.[39] Posner and Weisbach argue that there are important trade-offs between climate reduction and other goals, like tackling infectious diseases.

We might respond in at least two ways. The first is to question the certainty on which these cost-benefit analyses are built. While I do not doubt these can (and are) done, their reliability on matters of climate change is very limited. We need to be able to count the costs of environmental damage, perhaps including species extinction. But if such things are non-compensatory goods (as discussed in Chapter 1), putting a price on them is speculative at best and undermines the integrity of the modelling. And even if this were not so, economists have struggled to predict markets in recent years such as the 2008 financial crisis and the 2020 coronavirus pandemic. If it is so difficult to predict the next few years, the next century or more seems much more like guesswork. And the costs of getting decisions wrong could tip us all into environmental catastrophe.

A second response is to say that the cost-benefit approach is more value-laden than scientific. What we value makes a difference for how we assess what is a cost or benefit, but also how we balance them. For example, Posner and Weisbach claim there are trade-offs between focusing solely on climate reduction versus goals like tackling infectious diseases. They suggest it is more cost effective to reach annual incomes of $3,000 or more and so remove the risks of contracting malaria, if they are correct, than to only tackle climate change. But it is a value judgement whether the risks for some in poorer countries contracting malaria for a while longer is a price worth paying for ensuring that no one is at greater risk of climate change's harmful effects.

In reply, Weisbach might say we should honor "the iron law of wealth" which is that "increased wealth means increased energy use. . . . No nation, regardless of its political system, culture, or fantastic environmental values, has discovered a way to be wealthy without energy use."[40] Our relative prosperity depends on a global economic system that is substantially driven by fossil fuels. Making significant cuts immediately risks creating serious problems for the immediate social and economic prospects for many.[41]

Of course, we are living at such a moment now in the wake of the coronavirus pandemic whose full economic impact is yet to be seen. But that this "iron law of wealth" is a rule worthy of continued respect or we should choose to fundamentally transform the global economic system so that the link to fossil fuels is less or even broken is again a statement of values. Whether we are persuaded may depend on our faith in markets and the predictive power of counting the costs of damage to environmental goods beyond value.

Risks and precaution

This brings us to the understanding of relevant risks. Adaptation proponents are less risk-averse than mitigation advocates. If no measures whether adaptive or pro-mitigation are taken, then there is a genuine risk of our reaching dangerous tipping points and environmental catastrophe. Not only do adaptation proponents believe it possible to satisfactorily adapt, but they also have greater skepticism about the likely danger of reaching tipping points in the foreseeable future. We do not know what this point is – or how close we are to it today.[42]

This is perhaps more than an interpretive debate over the role of the *precautionary principle* because of the size of the relevant risks.[43] Adaptation may yield beneficial improvements in the short-term, but be reckless if directing a long-term policy. The adaptation approach has a more casual concern about the moral permissibility of exposing the environment to greater risk than conservation.[44] All living things require resources and impact the environment. The question is whether it is morally permissible to expose higher than necessary risks where it might be avoided. Adaptation proponents must argue more persuasively for why these risks are morally permissible and not merely more cost-effective.

Consider the idea that climate change creates harms, including the serious risk of harm. John Stuart Mill's *harm principle* states: "[t]hat the only purpose for which power can be rightfully exercised over any member of a civilized community, against his will, is to prevent harm to others. His own good, either physical or moral, is not a sufficient warrant."[45] How we understand what can count – and what does not count – as "a harm" is central to applying this principle. Lightning can harm, even kill, but is not relevant as the kind of harm that Mill has in mind. Instead, harms might include extreme weather patterns, droughts or the spread of tropical diseases to new areas made possible through creating climate changes.

But we stand at a possible philosophical fork in the road. If human activities contribute to harms, or risk of harms, then these activities must clearly cease – and those who harm might even be compelled to stop. We might even require compensation. But does this require reducing emissions? Possibly not.

We can compensate others through reparations, for instance, if harmed by our greenhouse gas emissions. Or we might compensate by enabling others to become unaffected by the relevant climate changes. Rising sea levels threatening coastlines could be compensated by moving those affected to new housing or building new flood defenses. The risk of harm can be removed so that no harm is imposed.

A possible reply is that adaptation does not expose others to risk, but rather helps create new conditions which foreclose risks from arising in the first place.[46] We do not then harm others through exposure to risk first, but remove risk later – instead, we remove this risk altogether.[47] This reply rests on high confidence that such risks may be identified early and removed before they become threatening. We need further evidence to justify such confidence over such speculative matters – and I am doubtful of its possibility.

Conclusion

Adaptation is an important aspect of any viable climate change policy. The climate is already changing and we must change with it.

The problem is whether any such general climate policy should primarily focus on adaptation instead of mitigation. There are many reasons to doubt that any such policy would be wise. There is too much faith on untested and unknown technologies that may have counterproductive outcomes in an uncertain future. The need for adaptive technologies is compelling as part of a broader strategy, but it cannot serve as the primary focus. No deep appreciation of the precautionary principle is needed to recognize the high risks at stake in misplacing our belief in technology not yet invented to save our planet from climate change's future effects. There is no evidence that adaptation will or is very likely to create the happily ever-after its proponents promote.

Nonetheless, some combination of a "*massive growth* in energy efficiency and clean renewable energy investments" alongside slashing carbon emissions and reducing consumption will be a part of the solution – although this "solution" to the problem of climate change remains elusive up to this point.[48] We must now consider an alternative perspective after having clarified the limitations for both mitigation and adaptation strategies.

Notes

1 U.N. Environment Programme, "Climate Adaptation" (2020), url: www.unep.org/explore-topics/climate-change/what-we-do/climate-adaptation. For further information about the UN's Global Adaptation Network, see www.unenvironment.org/gan/.
2 See Giddens, *The Politics of Climate Change*, 161: "Whatever happens from now on, climate change is going to affect our lives and we will have to adapt to its consequences." See also Dale Jamieson, *Reason in a Dark Time: Why the Struggle against Climate Change Failed–and What It Means for Our Future* (Oxford: Oxford University Press, 2014).
3 See UN Environment Programme, "Facts about the Climate Emergency."
4 See Mastrandrea and Schneider, *Preparing for Climate Change*, 13.
5 Stern, *A Blueprint for a Safer Planet*, 58.
6 See John Shepherd et al., *Geoengineering the Climate: Science, Governance and Uncertainty* (London: Royal Society, 2009): ix, 1.
7 Gardiner, "Ethics and Global Climate Change," 573.
8 See Aryn Baker, "Can a 4,815-Mile Wall of Trees Help Curb Climate Change in Africa?" *Time* (September 12, 2019), url: https://time.com/5669033/great-green-wall-africa/.
9 See Paul J. Crutzen, "Albedo Enhancement by Stratospheric Sulfur Injections: A Contribution to Resolve a Policy Dilemma?" *Climatic Change* 77 (2006): 211–219.
10 See Gardiner, *A Perfect Moral Storm*, 348.
11 This issue raises the question: what if an adaptive strategy – like sulfate infusion changing the color of the skies – did not have unanimous support? Given the gravity and scale it is conceivable for strong arguments to be made on both sides. If this is a matter of truly staving off catastrophe, this would weigh in favor of doing even if some countries objected. I favor a global unanimous consensus as the only means to legitimate such an intervention *politically*, but questions would need to be answered about its ethical and environmental legitimacy.
12 In the voluminous literature on climate change ethics, sewage management receives virtually no mention despite its being central to any sustainable future. This is a gap that should be filled. It is unclear whether to mention the management of basic bodily functions is too "dirty" or uncomfortable for ethicists and

others to consider – but it should not be any longer. I very much hope this brief intervention will spur more work in this area in future. It is urgently needed.

13 See David Shimkus, "Addressing Climate Change Means Addressing the Global Sanitation Crisis," *International Institute for Sustainable Development* (October 31, 2017), url: https://sdg.iisd.org/commentary/guest-articles/addressing-climate-change-means-addressing-the-global-sanitation-crisis/.
14 See Michelle L. Lute, Shahzeen Z. Attari and Steven J. Sherman, "Don't Rush to Flush," *Journal of Environmental Psychology* 43 (2015): 105–111. This study also found those who always flush were less likely to sacrifice for the environment than those who flushed occasionally.
15 Victor Hugo, *The Works of Victor Hugo: Les Miserables*, vol. 7 (London: Jensen Society, 1907): 100.
16 See Bilsen Beler Baykal, "Recycling/Reusing Grey Water and Yellow Water: Motivations, Perspectives and Reflections into the Future," *Desalination and Water Treatment* 172 (2019): 212–223 and Jongkwan Park et al., "Organic Matter Composition of Manure and Its Potential Impact on Plant Growth," *Sustainability* 11 (2346) (2019): 1–12.
17 See Lomborg, *The Skeptical Environmentalist*, 11.
18 It is a curiosity then that the leading approaches to climate change are *anthropocentric*. One possible reason is to address the problem of human motivation. Citizens and political leaders may be more likely to address climate change if our focus is on the potential threat to human sustainability. Besides, the spotted owl cannot vote. Nevertheless, it may be surprising that environmental ethics largely focuses upon how the environment might best sustain human beings rather than the more "green" consideration of how human beings might better serve nature. This is an important issue I bracket here.
19 See S. Matthew Liao, Anders Sandberg and Rebecca Roache, "Human Engineering and Climate Change," *Ethics, Policy and Environment* 15 (2012): 206–221.
20 See Raino Malnes, "Climate Science and the Way We Ought to Think About Danger," *Environmental Politics* 17 (2008): 660–672.
21 See S. Pacala and R. Socolow, "Stabilization Wedges: Solving the Climate Problem for the Next 50 Years with Current Technologies," *Science* 305 (2004): 968–972.
22 See R. Stuart Haszeldine, "Carbon Capture and Storage: How Green Can Black Be?" *Science* 325 (2009): 1647–1652. See also Elena V. McLean and Tatyana Plaksina, "The Political Economy of Carbon Capture and Storage Technology Adoption," *Global Environmental Politics* 19 (2019): 127–148.
23 See John Fogarty and Michael McCally, "Health and Safety Risks of Carbon Capture and Storage," *Journal of the American Medical Association* 303 (2010): 67–68 and Mark Z. Jacobson, "The Health and Climate Impacts of Carbon Capture and Direct Air Capture," *Energy and Environmental Science* 12 (2019): 3567–3574.
24 See Jorge H. Garcia and Asbjørn Torvanger, "Carbon Leakage from Geological Storage Sites: Implications for Carbon Trading," *Energy Policy* 127 (2019): 320–329.
25 Kahn, *Climatopolis*, 243.
26 Levitt and Dubner, *Superfreakonomics*, 177. See Lomborg, *Cool It*, 116–123.
27 Dale Jamieson, *Ethics and the Environment: An Introduction* (Cambridge: Cambridge University Press, 2008): 13.
28 Adaptation measures assume we will have sufficient resources to enable us to adapt to future climate changes. Suppose this were true. Such an approach may

be an available option, but prove irrational because long-term costs of adaptation may be much higher than conservation.

29 See Paul Wapner and John Willoughby, "The Irony of Environmentalism: The Ecological Futility but Political Necessity of Lifestyle Change," *Ethics and International Affairs* 19 (2005): 77–89.

30 Richard Wilkinson and Kate Pickett, *The Spirit Level: Why Equality is Better for Everyone*, revised ed. (London: Penguin, 2010): 223.

31 See Gardiner, *A Perfect Moral Storm*, 65.

32 Matthew E. Kahn, *Climatopolis: How Our Cities Will Thrive in the Hotter Future* (New York: Basic Books, 2010): 7, 12.

33 See Steven D. Levitt and Stephen J. Dubner, *Superfreakonomics* (London: Penguin, 2010): 169 (on "worldwide catastrophe" from climate change: "its likelihood is so uncertain").

34 See Stewart Brand, *Whole Earth Discipline*, revised ed. (London: Atlantic Books, 2010). On nuclear policy, see Thom Brooks, "After Fukushima Daiichi: New Global Institutions for Improved Nuclear Power Policy," *Ethics, Policy and Environment* 15 (2012): 63–69.

35 Lomborg, *The Skeptical Environmentalist*, 318.

36 See Stern, *A Blueprint for a Safer Planet*, 54.

37 See Lomborg, *Cool It*, 8, 35.

38 One related issue is whether economic theory can successfully address a cost-benefit analysis concerning climate change. See John Broome, *Counting the Cost of Global Warming* (Isle of Harris: White Horse Press, 1992): 19. This may be particularly true with regard to ascertaining the relevant costs to nonhumans and noneconomic costs to humans. See Mark Sagoff, *The Economy of the Earth* (Cambridge: Cambridge University Press, 1988) and David Schmidtz, "A Place for Cost-Benefit Analysis," *Noûs* 11(supplement) (2001): 148–171.

39 See Posner and Weisbach, *Climate Change Justice*, 29.

40 Weisbach, "Climate Policy and Self-Interest," 181. See K. Bithias and P. Kalimeris, "Re-Estimating the Decoupling Effect: Is There an Actual Transition toward a Less Energy-Intensive Economy?" *Energy* 51 (2013): 78–84.

41 See Gardiner, *A Perfect Moral Storm*, 20.

42 See Posner and Weisbach, *Climate Change Justice*, 54.

43 See Stephen F. Haller, *Apocalypse Soon? Wagering on Warnings of Global Catastrophe* (Montreal: McGill-Queens, 2002) and Cass R. Sunstein, *Laws of Fear: Beyond the Precautionary Principle* (Cambridge: Cambridge University Press, 2005).

44 For example, see Astrid Dannenberg and Sonja Zitzelsberger, "Climate Experts' Views on Geoengineering Depend on Their Beliefs about Climate Change Impacts," *Nature Climate Change* 9 (2019): 769–775.

45 John Stuart Mill, *On Liberty* (Indianapolis: Hackett, 1978): 9.

46 For example, see Thom Brooks, "Climate Change and Negative Duties," *Politics* 32 (2012): 1–9.

47 My thanks to Lea Ypi for pushing me on this point.

48 Robert Pollin, "The Green Growth Path to Stabilization," in Akeel Bilgrami (ed.), *Nature and Value* (New York: Columbia University Press, 2020): 117–126, at p. 123.

4 Climate change and catastrophe

"You must do the thing you think you cannot do."

– *Eleanor Roosevelt*

The need for a new perspective

Our challenge is not to determine whether there is climate change or its associated dangers, but rather how best to respond to it. This challenge does not admit of ready answers as the issue is complex and difficult. Much of my discussion in the preceding chapters have focused on where different approaches have proven unsatisfactory by their own lights.

While most commentators endorse some combination of mitigation and adaptation measures, their proposals usually emphasize one over the other. I have argued that neither conservationist proposals such as the ecological footprint or the polluter pays principle nor adaptation strategies are convincing individual solutions to the problem of climate change.[1]

This is important because each has adherents who claim that only if we implemented the footprint approach, polluter pays principle or an adaptive strategy, we could "solve" the problem of climate change by bringing the impact of such changes to an end. If this is a goal, these approaches are unsuccessful and we require a new approach or perspective.[2]

Climate change has no end-state solution

These strategies share a common mistake concerning the nature of the central problem. Both conservation and adaptation proponents claim their approach provides a *solution* to the problems associated with climate change.[3] Conservationists argue that adopting a policy based around ecological footprints or a polluter pays principle will lead to a sustainable future. Likewise, adaptation proponents claim we should focus our efforts on adapting to future climate change along with modest reductions in carbon emissions to ensure a sustainable future and even "a happy ending."[4] Most defenders of each view regularly conceive of a "solution" as a permanent state of affairs.

Both approaches offer what we might call *an end-state solution* to the problem of climate change: "The world now has the technologies and financial resources to stabilize climate."[5] The possibility of permanently ending climate change is conceived as an achievable end-state. If only governments followed the correct approach, then the problems arising from climate change could be solved permanently.[6]

The belief in an end-state solution to climate change is based on a fundamental mistake about its problem. Jonathan Schell makes the point eloquently:

> In this world without us, traditional evolution would indeed revive, and the procession of geological ages would resume, though without anyone around to give them names (unless in the fullness of time a new creature, perhaps some gifted fish, evolves to the point at which it can assume the task.[7]

Global environmental catastrophes have happened at least five times in our planet's past. Ice ages, major volcanic eruptions and meteors striking the Earth have each caused mass extinctions millions of years before human beings first evolved. These destructive periods have names: the Ordovician, Permian, Triassic and Cretaceous periods.[8] They are also well-known. For example, every schoolboy or schoolgirl learns that dinosaurs dominated the Earth until their mass extinction and without ever having come into contact once with human beings.

These facts about our planet make clear that mass extinctions and environmental catastrophes do not require human activity – they have happened to deadly effect several times. At least one period of mass extinction – the Cretaceous period – is believed to have been caused by a major meteor strike. If correct, this further suggests that environmental catastrophe can strike us no matter how climate-friendly our ecological impact.

Therefore, it is a mistake to theorize any mitigation, adaptation or combination thereof strategy that is likely or can guarantee a sustainable planet for all and a happy ever-after. We can no more stop the climate from ever changing than we can prevent the world from turning. Seeking a solution that can overcome this is doomed to failure. We need to retheorize what we want climate change ethics to achieve. Current end-state solutions lull us into a trap of false hope.

New horizons

The problem of climate change is not that it changes constantly; it is that its changes are more rapid – due primarily to human causes – rendering it more difficult and complex. For example, take our rising sea levels. These are a significant difficulty threatening millions of people worldwide. But let us be clear about the problem. Sea levels have been far higher in the distant past. About three million years ago the sea level was 10 to 40 meters (about 32 to 130 feet) higher than today – long before human beings walked the Earth. In only a matter of decades, our carbon emissions have raised the CO_2 in the atmosphere to a level not seen since that time.

Climate change happens. Its effects can be significant whether or not there are any human beings remaining to experience this. This changes our perspective horizon. We live in an endangered world. Environmental catastrophes can and have happened. If we left no impact, this would not ensure that catastrophes would be avoided in future. We live under a shadow.

It is a grave mistake to think that because the climate might change regardless of human activity that our actions lack consequences. In the Introduction, the global scientific consensus published in Intergovernmental Panel of Climate Change reports was clear and beyond doubt The primary driver today of climate change and its damaging effects is human activity. We bear a heavy responsibility for our actions and their effects on our climate.

So our new horizon is the impossibility of preventing any climate changes. A perfect principle for mitigation or advanced adaptive technology will not stop changes from ever happening. An end-state solution is beyond our grasp.

This raises a fundamental issue about our shared future. Let us first consider a view we should reject – and be clear why we should reject it. This is the thought that if the climate might change regardless of human activity, including the possibility of catastrophe through some natural or intergalactic event, then our efforts at promoting a sustainable future are futile. We cannot promise a happy ever-after without catastrophe. So, we might be tempted to believe, we should live each day as if it is our last. Who knows? It could be. We cannot guarantee a catastrophe-free future for ourselves, our children or future generations. Our focus can only be on the here and now. Or so the claim might be formulated. It is a mistake and it requires us to retheorize sustainability in our present context.

Retheorizing sustainability

The fact that our climate will change is not a compelling reason to act as if we are powerless. Our actions have consequences for our changing climate. The most notable effect is that our actions are hastening the arrival of potentially catastrophic circumstances. If we cannot forever prevent any environmental catastrophe leading to mass extinction of plant and animal life as well as quite possibly human lives, we should do everything possible to make such an event no more likely than we can. This gives us every reason to make full use of pursuing mitigating and adaptive measures – but as a means of not hastening, rather than preventing, any future catastrophe.[9]

We should approach climate change with a new understanding of sustainability in the context of our endangered world. The term "sustainability" carries many connotations, including safe, secure, maintain and support. When we normally speak of sustainable activities we think of what is able to be safely maintained; activities that are continuous and worth supporting.

This traditional view of sustainability does not neatly apply to climate change. A permanent state of total safety, security and support are not within our ability to deliver. Call this the *orthodox view*. The pursuit of a sustainable future must mean something else. It must be in light of a possible future environmental catastrophe that is both foreseeable and perhaps inevitable.

I propose we retheorize how we think about "sustainability" so that it better fits the world we find ourselves in. Our pursuit of sustainability is a seeking of continuation and security against the vulnerability that comes with impermanence. Call this alternative perspective *impermanent sustainability*.

Let us compare these different conceptions. An orthodox view of sustainability would motivate individuals to act safe in the knowledge that adopting the right proposal for addressing climate change is available and permanent. Sustainability means achieving a deliverable, fixed, forever *status quo*. Maintaining sustainability requires commitment, but the more difficult tasks are identifying what is required and embedding the right habits. For example, an orthodox view of the ecological footprint is we consider what limits its acceptance imposes on our consumption – and then we adjust our lives within those fixed boundaries.

Once we adopt the appropriate behaviors this view of sustainability is about maintaining these activities. Sustainability is threatened if we departed from this (relatively) fixed plan, this *status quo*. Of course, proponents of ideas like an ecological footprint would regularly make adjustments to its size in relation to changes in global population, resource needs and technological capabilities, for example. However, the model would only be tweaked, not recalculated so that it would lead to any significantly different effects unless there was some significant change in population, resource needs or new technological advances over a short time.

After we have found a workable way of calculating our footprint, our main task is to stick to our formula – and the climate would stabilize around a new normal, or *status quo*. Sustainability is achieved through climate stability. There is an end-state solution to our problem. Greenhouse gas emissions are not a serious problem once we reset our lives within the boundaries of our share of ecological space.

Impermanent sustainability is a very different working model. It lacks a fixed plan and so continually readjusts its plans anew. There is no new *status quo* to reach. The future is not a given and end-state solutions are a fantasy. This requires our continuous commitment to our own survival seeking the best evidence and reducing our impact however possible. For example, it rejects the orthodox understanding of what an ecological footprint can achieve. Impermanent sustainability denies that there is any fixed model or formula for the future. There is no button we can press that stops the climate from changing.

Sustainability is not finding a set way of living, but *constant reinvention.* Any emissions or other impact are a source of potential problems. Our world is nonideal. Even if we constrained our consumption to much less than any footprint leaving only a trace imprint on the environment, this could be sufficient to trigger a catastrophe if our climate was on the edge of falling over a tipping point – whether as part of a natural cycle or accelerated by our historical consumption and emissions. Our actions might not prevent a catastrophe, but they can make it more or less likely – and more or less dangerous. (I return to this point in the next section.)

One way of conceiving the differences between the orthodox view of sustainability and this new view of impermanent sustainability is this: it is disruptive for us to achieve the former, but continually disruptive for us to maintain the latter. The shift in our lifestyles to meet the mitigation goals of the ecological footprint or polluter pays principle would be transformative. Our ways of being and doing would be changed. But the primary disruption is in making the shift. In contrast, the fundamental changes required by the impermanent view of sustainability are disruptive to our current "business as usual" and may require further shifts in the future depending on new adaptive technologies and our collective ability to conserve more – and better.

Challenging conventional thinking

Let us consider three important implications for the view that we should reject the search for end-state solutions to climate change and adopt a view of impermanent sustainability.

The first implication is it accepts a more risk adverse view of future climate changes. Instead of being satisfied with leaving an impact under a permanently "safe" level of consumption, impermanent sustainability seeks to reduce consumption and environmental impact as much as possible. Since it rejects any fixed end-state future, our ability to impact our climate a bit less will help us more in not accelerating the onset of an environmental catastrophe and buy us more time to ready for ourselves for the day it is triggered by natural or human causes.

My view is not that we should do all we can – even if falling short – because of pessimism that our communities and political leaders lack the commitment or resources to do everything possible to combat climate change. Instead, my position is realist and pragmatic: our doing everything possible does not render climate change impossible.

Learning to live with climate change is not resignation or a licence for inaction, but it means we must work even harder to ensure as sustainable a climate as we can. So we must ask new ethical questions like: How should we theorize impermanent sustainability and its constant reinvention? How to measure our progress against a constantly impermanent target? Are there limits to our interventions? And, if climate change is ineliminable, how do we measure success? A new climate change ethics for our endangered world raises questions like these.

Conventional work does not answer these questions. Doing what we can in terms of living under an ecological footprint, implementing carbon taxes and investing in adaptative technologies are ways of reducing the probability of catastrophe. But each usually presupposes that, if only that approach were followed, climate change might be stopped – which is a mistake. Even if each did not make that error, none are directly engaged with the concept of impermanent sustainability which characterizes climate change. Nor do they help us grapple with defining success, especially when advocated piecemeal and not as a part of a larger, more comprehensive project. We must turn to a different perspective.

A possible objection is our climate might have fluctuated in the past, but this pattern might not be repeated. It might be thought that my view of sustainability assumes too much in expecting such an event in future, or so it could be argued.

There is much less evidence for the view no catastrophe is likely or within our ability to avoid in future. It is more science fiction than fact to believe we will ever be able to control the global environment with any confidence. Scientists may not be divided about whether the climate is changing, but there are serious divisions about its scale and even its nature.[10] For example, there is widespread debate about the most promising model for capturing climate change.[11] Climate scientists accept cyclical climate changes, but there is no clearly dominant view about their causes.[12] As one scientist puts it, "many aspects of ice-age dynamics remain a mystery."[13] In summary, we might know more about how to destroy our planet – through triggering environmental catastrophes or nuclear warfare – than how it might be preserved, or how to save it from ourselves. If so, it should be more compelling to accept a model that best accounts for a world subject to inevitable future catastrophe than not.

A second implication is impermanent sustainability requires we can remain motivated to continually mitigate our impacts and adapt ourselves under the shadow of an unavoidable future catastrophe.

Naturally, a possible objection to this view is the traditional challenge looms so large already that it may overwhelm us into a dangerous sense of becoming resigned to a catastrophic fate. Increasingly urgent, if not drastic, measures are required to slow the worryingly deep impact our consumption is having. If I am right, then the new model of impermanent sustainability will fuel a stronger sense of hopelessness. Many will wonder what is the point of making continual changes to their daily lives if that extra effort does not end the possibility of an environmental catastrophe in future.

Such pessimism should give way to realism – where there is hope. It is a mistake to believe our efforts would be in vain if a future catastrophe remained possible. This is because our activities can make this happen more quickly by accelerating climate change through our impact. We should do everything we can to delay such a devastating event – and what we do can make a significant difference.

Our activities can also make any such catastrophe no more damaging than it might be otherwise. If we continue to accelerate climate change to an unprecedented degree, we move forward the possibility of catastrophe with less time to prepare for it and we put rocket boosters on its impact in pushing our climate so hard – like making a tidal wave heading for our coastline larger while approaching faster. Such a reaction would make such a disaster even more difficult to endure.

We can work together swiftly and comprehensively, as the current global Covid-19 pandemic makes clear. The world has entered an unprecedented lockdown to tackle this emergency. Previously unimaginable cuts in carbon emissions have been achieved within days as hundreds of millions of people stopped travelling to work. The Centre for Research on Energy and Clean Air estimated the climate benefits from the lockdown in the UK and across Europe led to 11,000 fewer deaths with a fall of 40 percent or more in major pollutants in countries like Italy, Spain and France. While there has been a steep rise in unemployment, there was also 1.3 million fewer days of work absence.[14] European emissions have been cut by an estimated 390 million tons – China has seen a reduction of 250 million tons.[15] Major change is possible for the better if we want it. This should help motivate us to work across borders, to think big, and our collective full potential is far from tapped.

Any change is welcome. Reducing our impact to zero is best. But this will not prevent any possibility of catastrophe from happening – although it will delay it or lessen its strength had our impact been greater. So claiming that serious measures must be enacted to avoid a future catastrophe is mistaken, but taking the most substantial measures possible will bring significantly positive results. However, failing that, any progress makes some difference. We need to motivate ourselves by doing what we can and getting into continually better habits. Something is much better than nothing.

Conclusion

It is not unique to claim that our theories seem inadequate for dealing with longer term problems in general.[16] Previous chapters considered how and why mitigation and adaptation approaches claim to deliver long-term solutions. It was argued that these have a crucially important place in any climate change strategy, but no one proposal could or would deliver the sustainable forever its defenders have promised.

This chapter has gone further in saying these approaches are built on a mistaken assumption about the problem they seek to "solve." We should not look for an end-state solution because our climate is changing with no off button to make it stop. Therefore, the orthodox view of sustainability as stability subject to our adopting different lifestyles and technologies does not fit our reality.

Instead, we should view climate change as a *challenge*. The fact of change is not a problem insofar as it cannot be solved by stopping it. However, the effects of climate change do pose problems for how we adapt ourselves to an ever-changing climate. I have advocated a new model of impermanent sustainability to guide our thinking. We should be driven by the need to continually revaluate ways of reducing our impact notwithstanding if it is already within some defined ecological footprint or not. We should not be overconfident in what future technology can deliver – but we should buy ourselves time to make as much progress as necessary by preparing ourselves for the inevitability of environmental catastrophe whether it is ultimately caused by our activities (which appears certain) or through natural causes like volcanic activity and meteor strikes (as has happened before).

I strongly agree with Dale Jamieson's warning: "do not let the perfect be the enemy of the good."[17] We live in a nonideal situation. The importance of conservationist and adaptation strategies does not disappear because they might only manage, but not "solve" the problem of avoiding environmental catastrophe. My point is that these approaches should be considered in a new light – and are urgently required.[18]

Nor should we underestimate the powerful inborn drive for survival within us all. We can make significant forward progress if we are committed to it, as seen in reactions to the current global pandemic – unthinkable only a few months ago. Acting as if it no longer matters will only accelerate the worst excesses of a future catastrophe. We can – and we must – do better.

Notes

1 See Gardiner, *A Perfect Moral Storm*, 7: "existing theories are extremely underdeveloped in many of the relevant areas, including intergenerational ethics,

international justice, scientific uncertainty, and the human relationship to animals and the rest of nature."

2 To be clear, there are supporters of mitigation and adaptation measures who have more modest ambitions and who are explicit in saying their preferred approaches do not offer a longer term solution or strategy to climate change. However, they are a minority. I do not deny the fact that most commentators taking either approach might accept additional policies beyond what they recommend. However, it remains the case that most expressly claim to "solve" the problem of climate change if we adopt a favored proposal – *or* perhaps they might say something more is required, but leave this unacknowledged. This is not a question of making clearer the combination of mitigation and adaptation approaches within a coherent policy aimed at better managing climate change, but instead the failure of most commentators to acknowledge the limits of favored policies as an end-state view that might do no better than temporarily manage the climate change we experience. For one fine exception, see Jamieson, *Reason in a Dark Time*.

3 See Thom Brooks, "The Real Challenge of Climate Change," *PS: Political Science and Politics* 46 (2013): 34–36; Thom Brooks, "The Inevitability of Climate Change," *Global Policy* 5 (2014): 112–113 and Thom Brooks, "Why Save the Planet?" in Thom Brooks (ed.), *Current Controversies in Political Philosophy* (London: Routledge, 2015): 138–147.

4 Kahn, *Climatopolis*, 12.

5 Lester R. Brown, *World on the Edge: How to Prevent Environmental and Economic Collapse* (New York: W. W. Norton, 2011): 198. See Thom Brooks, "Climate Change Ethics and the Problem of End-State Solutions," in Thom Brooks (ed.), *The Oxford Handbook of Global Justice* (Oxford: Oxford University Press, 2020): 241–258. The concept of "end-state" results is from Robert Nozick, *Anarchy, State and Utopia* (New York: Basic Books, 1974): 153–155. See also M. C. Henberg, "Nozick and Rawls on Historical versus End-State Distribution," *South western Journal of Philosophy* 8 (1977): 77–84 and Biung-Ghi Ju and Juan D. Moreno-Ternero, "Entitlement Theory of Justice and End-State Fairness in the Allocation of Goods," *Economics and Philosophy* 34 (2018): 317–341.

6 One possible objection is that my characterization is inaccurate. Conservationists and adaptation proponents do not always expressly indicate that if we endorse a favored policy then climate change can be managed successfully without additional policies needed. I do not deny the fact most commentators taking either approach might accept additional politics may be required beyond what they recommend. It remains the case that several expressly claim to "solve" the problem of climate change if we adopt a favored proposal and I have highlighted several examples in this article. My critique refers to the general character of most work falling under either conservationist or adaptation approaches: if something more beyond adoption of an ecological footprint or polluter pays principle is required, then this is too often unacknowledged. This is not a question of making clearer the combination of conservationist and adaptation approaches within a coherent policy aimed at better managing climate change, but instead the failure of most commentators to acknowledge the limits of favored policies as an end-state view that might do no better than temporarily manage the climate change we experience. For one exception, see Jamieson, *Reason in a Dark Time*.

7 Schell, "Nature and Value," 5.

8 See David M. Raup and J. John Sepkoski, Jr., "Mass Extinctions in the Marine Fossil Record," *Science* 215 (1982): 1501–1503.

9 While there have been longstanding concerns about the possible inevitability of a future ice age, the planet's slow cooling is being trumped by warming caused by the increased greenhouse effect. While an ice age might now be much less likely than previously thought, now the concern has shifted to whether short-term catastrophe due to global warming is inevitable. So while the nature of the form any future catastrophe might take has shifted, there is no less a concern about the likelihood of an environmental catastrophe because of climate change and its possible effects.
10 See Richard A. Posner, *Catastrophe: Risk and Response* (Oxford: Oxford University Press, 2004): 43–58.
11 See Pascale Braconnot, Sandy P. Harrison, Masa Kageyama, Patrick J. Bartlein, Valerie Masson-Delmotte, Ayako Abe-Ouchi, Bette Otto-Bliesner and Yan Zhao, "Evaluation of Climate Models using Palaeoclimatic Data," *Nature Climate Change* 2 (2012): 417–424.
12 For example, see on glacial cycles: Ayako Abe-Ouchi, Fuyuki Saito, Kenki Mawamura, Maureen E. Raymo, Jun'ichi Okuno, Kunio and Takashi and Heinz Blatter, "Insolation-Driven 100,000-Year Glacial Cycles and Hysteresis of Ice-sheet Volume," *Nature* 500 (2013): 190–193.
13 Shawn J. Marshall, "Climate Science: Solution Proposed for Ice-Age Mystery," *Nature* 500 (2013): 159–160.
14 See Centre for Research on Energy and Clean Air, "11,000 Air Pollution-Related Deaths avoided in Europe as Coal, Oil Consumption Plummet," (April 30, 2020), url: https://energyandcleanair.org/air-pollution-deaths-avoided-in-europe-as-coal-oil-plummet/.
15 See Jonathan Watts, "Climate Crisis: In Coronavirus Lockdown, Nature Bounces Back–But for How Long?" *The Guardian* (April 9, 2020), url: www.theguardian.com/world/2020/apr/09/climate-crisis-amid-coronavirus-lockdown-nature-bounces-back-but-for-how-long.
16 See Gardiner, *A Perfect Moral Storm*, 41.
17 Jamieson, *Reason in a Dark Time*, 9.
18 See Jamieson, *Reason in a Dark Time*, 228.

5 Possible objections

> "*A nation that destroys its soil destroys itself.*"
>
> – *Franklin D. Roosevelt*

Reassessment

It is time to take stock of the argumentative narrative thus far. The Introduction confirmed the global scientific consensus that climate change is happening, primarily caused by human activities and requires urgent action. Chapters 1 and 2 explored mitigation strategies for reducing our greenhouse gas emissions through the idea of an ecological footprint, a fair share of the atmospheric sink and the polluter pays principle. While each makes important contributions to any climate policy in the short or longer term, no individual policy proposal provides that silver bullet delivering any permanent "solution" and so they do not live up to the high expectations of some vocal proponents.

Chapter 3 considered adaptation strategies for how technology might offer the "solution" that eluded mitigation approaches. Adaptation defenders are especially overconfident about what mostly untested or not yet invented advances will – and, indeed, must – achieve a future free from the negative effects of continuing climate changes. While adaptation is a necessary part – alongside mitigation – to any climate policy, the discussion up to that point was critical: there is no one view that can secure the happily ever-after that proponents promise. The right climate policy must adopt some new model or way of thinking.

Chapter 4 argued that – in addition to any conceptual problems in terms of how an ecological footprint achieves equity and fairness, among other examples – the main mitigation and adaptation strategies rest on a mistaken foundation searching for some means of enabling a permanent sense of sustainability. This gets wrong the nature of climate change as a problem that can be solved rather than this change an unstoppable element that we might slow or hasten, but not turn off. It was argued that we should retheorize sustainability as impermanent to match the current reality. This is a more demanding standard to meet the more difficult challenges ahead, such as the inevitability of environmental catastrophe.

In this chapter, we consider various possible objections to the argumentative narrative told thus far. These will be partly drawn from responses to my previously published work, but also extending much further to reflect on new potential concerns. The aim will be to clarify the critical positions taken and sharpen the analysis further.

The problem of motivation, revisited

Let us start by considering further a possible objection raised in Chapter 4. The concern might be explained like this. While agreed that I am right to focus on the inevitable changing nature of our climate and ever-present possibility of environmental catastrophe on our horizon, it is feared that I soften the urgency of our situation and miss what might be done to delay or blunt the worst outcomes.[1]

The concern that I soften the urgency of our situation is because if climate change is inevitable many people might be less inclined to take the significant steps we require to better prepare ourselves for any future catastrophe. A similar general worry is expressed by others, like Mike Hulme, who agree that climate change "will not be 'solved' by science or technology . . . [or] by politics or economics" because climate changes are inevitable.[2] But, instead, he says: "we need to ask not what we can do for climate change, but to ask what climate change can do for us."[3] He argues that we have more than two decades of Intergovernmental Panel on Climate Change reports based on a solid global scientific consensus about the dangers that climate change poses with too little action following afterwards.

Hulme suggests that in framing climate change as the "mother of all problems" we have created a monster ever more fearsome – and increasingly beyond our reach.[4] An example is in the belief that we mere human beings can control and dominate nature itself.[5] Rather than "fight" against climate change, we should mobilize ourselves in ways that situate our inability to control or forever prevent our climate from changing in more productive was, such as taking a more incremental focus.[6]

In reply, this is a disagreement primarily about tactics. I agree that harping on about the gravity of a problem has limited usefulness. While it can alert us to dangerously worrying events commanding our full attention, this sensitivity can wear off if it creates a sense of hopelessness in the face of inevitability. We need some reasons to be cheerful to motivate us.

Every one of us alive today will die. This inevitability can overwhelm some of us, but more often it helps us put extra meaning in life. We do not avoid relationships with others – we value them. Our environment is something precious. The inevitability of its change and future disasters can be used to inspire us to do what good we can – provided we have some means of positive action.[7]

The second, related concern is that in highlighting the inadequacies of different mitigating and adaptation strategies I neglect their importance for delaying or blunting a future environmental catastrophe. I do not intend any such neglect as, it is hoped, the previous chapter made clear. The inadequacies of mitigating and adaptation for providing a permanent solution is not meant to suggest we should discard them altogether. Imperfection does not mean rejection in our nonideal circumstances.

I accept that significant mitigation strategies slashing greenhouse gas emissions are central to any compelling climate policy. This can make the clearest long-term gains for any predictive modelling for how the next decades or centuries could play out. Carbon taxes akin to a polluter pays principle will almost certainly play a role in this as well. So, too, adaptive technologies. I agree that "the worst is not inevitable" if we can blunt or delay a future catastrophe.[8]

Theory and practice

Some may claim that my analysis confuses theory and practice. Philosophers Alexander Lee and Jordan Kincaid say that "the value of philosophy rests not on successful policy action, but in the process of moral evaluation."[9] Philosophy can continue to guide moral mitigation even in a world where mitigating the climate is no longer possible – and so climate ethics is immune from my critique, not least what is presented here.

In response, it is unclear what kind of guidance philosophy can bring where we cannot act on it, especially when we are grappling with *applied philosophy*, as we are here. If the polluter pays principle purports to provide an end-state solution which it cannot, in fact, secure, then perhaps there is ethical merit in its drawing attention to the importance of a negative duty that we have with the need to act on it. Nonetheless, we should not be unconcerned by the efficacy of proposals justified, at least in part, as a more just means to securing permanently sustainable mitigation. In applied philosophy, the application of philosophical contributions matters.

To be clear, it is *not* my view that other areas of philosophy are less important or do not matter. Logic, metaphysics, the philosophy of language and more all have highly valued contributions in virtually every area. Clarifying our concepts, their standing and any intended meanings and more matters, too. But this does not discount the importance of taking seriously how well the proposals from applied philosophy meet the real world challenges they set for themselves. If the polluter pays principle is unable to ensure all polluters pay for their pollution or if it is unable to have any plausible chance of securing sufficient mitigation *as it promises*, then our problem is with both its relevance for policy *and* "polluter pays" as a principle.[10]

How "new" is the new ethics?

A further possible objection is that the view of climate change ethics presented here is not entirely new. My criticisms of equal capita shares or the polluter pays principle needs not hit the mark as they could be advocated without a view to any end-state or happy ending. Furthermore, it is enough to say that we need to reduce emissions enough to reduce the probability of catastrophe, and to make it less damaging. The only questions are about how much each country should reduce its emissions by – and both ecological footprint and polluter pays principle proponents have answers to that question. Therefore, it seems forced to say both should be rejected because each accepts an end-state or happy ending view.

In reply, the earlier chapters are clear that I do not believe all advocates of views like the ecological footprint or polluter pays principle accept the position that, if only their approach was enacted, these approaches would necessarily lead to no further climate change and so a happy ending – even though I am also clear that many, if not most, of these proponents do hold this view. So this objection gets right that neither view need accept the happy ending conclusion, which I have accepted.

The real issue then is how each conceives of sustainability and progress towards achieving it. This is where each fall down most. I have raised numerous objections on grounds of fairness and equality to the kind of anthropocentric carving out of ecological space defended by proponents of an ecological footprint. I have also raised serious concerns with the calculation of polluter pays principles and what it might achieve.

But putting aside these criticisms, it is clear that even if these are put aside each approach is only partial. By themselves, they get us only so far. A more comprehensive strategy is more compelling – such as to also accommodate additional measures, including adaptative technologies – and compatible.

Therefore, it is not enough to say that any one approach can help justify reducing emissions and so reducing the probability of catastrophe – and so become the approach by which we address all questions relating to climate change ethics. If two or more approaches are compatible and more likely to reduce the probability of catastrophe even more, then such a more comprehensive strategy must be our focus instead – to be informed by an impermanent view of sustainability and a new way of understanding progress. So the view of climate change ethics here is different and so too its perspective.[11]

Partial solutions

Jonathan Peter Schwartz is concerned that my criticisms of the ecological footprint miss a key point about it. He agrees that affluent states are living beyond their means, and poorer states suffer from the change-related harms their overconsumption creates.[12] His concern is that my rejection of the ecological footprint as a *permanent* solution should not entail its rejection as a *partial* – and necessary part – of any compelling climate policy. Schwartz says – and I agree – that an ecological footprint makes possible a sustainable global economy that would help buy us time to pursue the adaptation strategy that I have advocated. So I should not throw out the baby with the bathwater, so to speak.

In reply, my criticism of the footprint strategy took more than one form. I argued its one-size-fits-all footprint does not treat countries equally (e.g., some will have greater or smaller resource needs depending on local climate) or fairly (e.g., some individuals over a lifetime will require differently sized footprints). I further observed that locking countries into the same-sized footprints relative to their populations would likely ossify the privileged positions of the affluent over developing countries, because the former would be best placed to exploit these conditions to their relative advantage.

So if our reason for supporting the footprint approach is a desire to improve conditions for the global poor, it is unclear that the global order will necessarily become more equal through use of the footprint model alone. On these specific grounds, I raised doubts about how adequately the ecological footprint fulfils its commitments to equality and fairness for all – and I rejected it as a permanent "solution."

But I have also tried to make clear throughout that while some advocates for the ecological footprint *not all do*. Moreover, *as a part* of any compelling strategy, the ecological footprint can play an important role – but not the only role. So I agree with Schwartz's position.

Overpopulation

One objection that I have not yet considered is whether the main challenge is overpopulation – and not our climate. It is true that the world's population is growing to an unprecedented size from 1.6 billion in 1900 to nearly 8 billion today. It is also true that this ever increasing population is putting significant pressure on resources, which can be limited and often non-renewable.

It is not a new argument to say, strictly speaking, that the problem of overpopulation is not about numbers, but consumption. The key question to ask concerns the use of resources. The size of the population will undoubtedly impact on that. However, different groups of people may make use of varying amounts of resources. It is this pressure on resources and their consumption (including emissions) that matters – and this can be managed in ways. A larger population need not always be a case of growth gone too far in terms of numbers if we can better accommodate their consumption needs. Consumption, not numbers, counts most.

There are real pressures on resources, especially those which are limited and non-renewable. Some argue their abundance leads to a *natural resources curse* promoting political instability and conflict.[13] But this seems to be a problem of how wealth is shared in those countries and a lack of constitutional safeguards.[14] One area of growing research is the field of space law, especially in relation to property rights as countries begin to consider using space exploration for mining resources on other planets after plundering what is left on our own planet. This opportunity remains some years off.[15]

Climate change as a violation of human rights

Why not a human rights-based (or "climate rights") approach? The case can appear compelling. There are multiple rights in widely supported international agreements and treaties that relate to the environment. One example is the International Covenant on Economic, Social and Cultural Rights adopted in 1966 that granted rights like to "freely dispose of their natural wealth and resources" and "in no case may a people be deprived of its means of subsistence," the right of everyone to "an adequate standard of living" including freedom from hunger and a responsibility to prevent, treat and control disease.[16] A second example is the 1948 U.N. Declaration of Human Rights which includes rights to "life, liberty and security of person."[17] Or the European Convention on Human Rights regarding the right to life and right to respect for private and family life.[18]

Our ability to use and dispose natural resources, our having access to means of subsistence and an adequate standard of living, to be free from hunger and disease and a right to life all connect to living within a suitable climate. The adverse effects of climate change caused by others seems able to infringe those rights – most of which we might consider to kinds of human rights with a need for clear protections. For these reasons it is easy to see why many can draw the conclusion that climate change can threaten, if not violate, our human rights.[19]

The interconnections between climate change and human rights law remains an evolving area.[20] Thus far, the heavy lifting – and courtroom results – seem driven not from the strength of the relevant human rights claims, but from what evidence there is that a state has not, or could not, meet a binding obligation to greenhouse gas emission reduction targets under international law.[21] Progress has been made in the courts in holding states to their international treaty commitments and not the breach of human rights. An impact of this is that the wrongs of climate change – as violating individual rights – gets lost. The problem of climate change becomes only a failure to meet contractual obligations. This seems to miss the point of the kinds of harms that climate change represents.

The main problem for establishing human rights claims in relation to climate change is meeting specific causation requirements like the so-called "but for" test.[22] This is a common legal rule to establish that some alleged rights infringement was caused by specific cause. Someone can be found to have breached my rights when we can establish that my right would not be infringed *but for* the actions of my transgressor.

This is a very difficult kind of test for climate change events to meet. It is not enough to show that the sea level has arisen or that someone has been adversely affected by an extreme weather event. Nor is it enough to connect such natural events and disasters to climate change more generally. We must link the actions of an agent (or agents) to contributing to climate change which caused harm to another.

Advancements are being made in pinpointing contributions. For example, Klaus Hasselman developed new methodologies to attribute observed trends in the global mean temperature to actors in 1997 through his "model of climate variability."[23] But while we might link climate change to a high statistical causality of specific climatic events, who is specifically to blame – and how might this responsibility be shared? Our knowledge does not appear precisely enough yet to establish in law such possibilities of climate change breaching human rights.

There is little doubt that such breaches are happening, only not yet meeting the evidentiary requirement for succeeding in court. Further work in modelling climate variability might improve such efforts. Perhaps we can do no better than be generally reassured. I accept Amartya Sen's claim that "it is better to be vaguely right than precisely wrong."[24] Relatedly, Nicholas Stern reminds us that: "we cannot afford to wait until we know everything with certainty."[25] If we cannot act until such breaches are substantially evidenced, but then by that point it becomes too late to act – we may find ourselves in a vicious circle.

We should look to some different standard. An attractive candidate for this role is the capabilities approach. It may be worth exploring the enforcement of a capability, not a right, to the environment.[26] A capability is the right to do or be – and they may often overlap with our human rights. Martha Nussbaum defines our capabilities as including our "being able to live with concern for and in relation to animals, plants and the world of nature."[27] Such a statement would require different countries to specify how this would be protected and enforced in their jurisdictions. But it seems to me our being able to establish my lack of being able to do or be with respect to the damaging effects of climate change a different kind of test. Either way, it seems clear our rights and freedoms are increasingly under threat from climate change. More must be done to protect this from happening further as our climate continues to change. Nussbaum's capabilities approach could be a way of better realizing this.

Conclusion

This chapter began with a stocktaking and summary of the overall argumentative strategy of the proceeding chapters. Several possible objections across a range of concerns and issues were explored to clarify and sharpen the positions defended in this book. They help us reaffirm that the way we view climate change and expectations for future policy must change. The climate is changing and we must change with it, while living in the shadow of an ever-present possibility of endangerment.

Notes

1 See Andrew Jameton, "Time Frames for Saving the Planet," *Ethics, Policy and Environment* 19 (2016): 136–140.
2 See Hulme, *Why We Disagree about Climate Change*, 329.
3 Hulme, *Why We Disagree about Climate Change*, 326.
4 See Hulme, *Why We Disagree about Climate Change*, 333.
5 See Hulme, *Why We Disagree about Climate Change*, 351.
6 See Hulme, *Why We Disagree about Climate Change*, 360.
7 See Clement Loo, "Environmental Justice as a Foundation for a Process-Based Framework for Adaptation and Mitigation: A Commentary on Brooks," *Ethics, Policy Environment* 19 (2016): 145–149; Ben Mylius, "Change-Oriented Conceptions of Climate: A Response to Thom Brooks' *How Not to Save the Planet*," *Ethics, Policy and Environment* 19 (2016): 150–152 and Eoin O'Neill, "The Precautionary Principle: A Preferred Approach for the Unknown," *Ethics, Policy and Environment* 19 (2016): 153–156.
8 See Jameton, "Time Frames for Saving the Planet," 139.
9 See Alexander Lee and Jordan Kincaid, "Two Problems of Climate Ethics: Can We Lose the Planet but Save Ourselves?" *Ethics, Policy and Environment* 19 (2016): 141–144, at p. 142.
10 See Lee and Kincaid, "Two Problems of Climate Ethics," 142.
11 I am deeply grateful to Peter Singer for suggesting this set of possible objections.
12 See Jonathan Peter Schwartz, "On Staying Focused: Response to Thom Brooks' *How Not to Save the Planet*," *Ethics, Policy and Environment* 19 (2016): 157–159.
13 See Leif Wenar, *Blood Oil: Tyrants, Violence and the Rules That Run the World* (Oxford: Oxford University Press, 2016).
14 See Thom Brooks, "The Resource Curse and the Separation of Powers," *Ethics and International Affairs* (blog) (April 2013), url: www.ethicsandinternationalaffairs.org/2013/the-resource-curse-and-the-separation-of-powers/.
15 The possibility of establishing interplanetary bases for mining – and distributing – natural resources from our solar system and beyond would be a further extension of debates about justice that moved from the state to global justice and then to *intergalactic* justice in outer space. There does seem the potential for this

technological advancement to yield significant benefits, but also very likely exacerbating existing inequalities. This is because the most affluent and dominating countries now are in the strongest position to take advantage of this opportunity and its benefits to the exclusion of states that are poorer and less powerful.

16 International Covenant on Economic, Social and Cultural Rights, New York, December 16, 1966, in force January 3, 1976; Articles 1, 11, 12(c).

17 UN Declaration of Human Rights, Paris, December 10, 1948; Article 3.

18 See European Convention on Human Rights; Articles 3 and 8. See also Ingrid Leijten, "Human Rights versus Insufficient Climate Change: The Urgenda Case," *Netherlands Quarterly of Human Rights* 37 (2019): 112–118.

19 See Simon Caney, "Human Rights, Climate Change and Discounting," *Environmental Politics* 17 (2008): 536–555.

20 See Alan Boyle, "Human Rights and the Environment: Where Next?" *European Journal of International Law* 23 (2012): 613–642; Lavanja Rajamani, "The Increasing Currency and Relevance of Rights-Based Perspectives in the International Negotiations on Climate Change," *Journal of Environmental Law* 22 (2010): 391–429 and Toby Svoboda, Holly Buck and Pablo Suarez, "Climate Engineering and Human Rights," *Environmental Politics* 28 (2019): 397–416.

21 See Alan Boyle, "Climate Change, the Paris Agreement and Human Rights," *International and Comparative Law Quarterly* 67 (2018): 759–777 and Petra Minnerop, "Integrating the 'Duty of Care' under the European Convention on Human Rights and the Science and Law of Climate Change: The Decision of the Hague Court of Appeal in the Urgenda Case," *Journal of Energy and Natural Resources Law* 37 (2019): 149–179.

22 For critical analyses of "but for" causation, see Jane Stapleton, "An 'Extended-But-For' Test for the Causal Relation in the Law of Obligations," *Oxford Journal of Legal Studies* 35 (2015): 697–726, James Edelman, "Unnecessary Causation," *Australian Law Review* 89 (2015): 20–30 and D. James Greiner, "Causal Inference in Civil Rights Litigation," *Harvard Law Review* 122 (2008): 533–598.

23 See Klaus Hasselmann, "Multi-Pattern Fingerprint Method for Detection and Attribution of Climate Change," *Climate Dynamics* 13 (1997): 601–611 and G. C. Hegerl, Klaus Hasselmann, U. Cubasch, J. F. B. Mitchell, E. Roeckner, R. Voss and J. Waszkewitz, "Multi-Fingerprint Detection and Attribution Analysis of Greenhouse Gas, Greenhouse Gas-plus-aerosol and Solar Forced Climate Change," *Climate Dynamics* 13 (1997): 613–634.

24 See Stephen M. Gardiner, "Betraying the Future," in Stephen M. Gardiner and David A. Weisbach (eds.), *Debating Climate Ethics* (Oxford: Oxford University Press, 2016): 6–45, at p. 39. See also See Eric Brandstedt and Anna-Karin Bergman, "Climate Rights: Feasible or Not?" *Environmental Politics* 22 (2013): 394–409.

25 Stern, *A Blueprint for a Safer Planet*, 6.

26 See Thom Brooks, "Respect for Nature: The Capabilities Approach," *Ethics, Policy and Environment* 14 (2011): 143–146. See also Thom Brooks, "The Capabilities Approach and Political Liberalism," in Thom Brooks and Martha C. Nussbaum (eds.), *Rawls's Political Liberalism* (New York: Columbia University Press, 2015): 139–173 and Thom Brooks, "Capabilities, Freedom and Severe Poverty," in Thom Brooks (ed.), *The Oxford Handbook of Global Justice* (Oxford: Oxford University Press, 2020): 199–213.

27 See Martha C. Nussbaum, *Women and Human Development: The Capabilities Approach* (Cambridge: Cambridge University Press, 2000): 34.

Conclusion

> *"The eyes of all future generations are upon you. And if you fail us, I say we will never forgive you."*
>
> *– Greta Thunberg*

A new ethics

Climate change is happening and accelerating. Its effects are growing. The primary cause is human activities like greenhouse gas emissions. There are no doubts about this global scientific consensus.

Doing nothing is not an option. The climate's changes would continue at an unprecedented pace – and our vulnerability to them would grow as well. Countries are becoming submerged by rising tides, ice caps and glaciers are melting, droughts are becoming more commonplace and intense, extreme weather conditions more frequent and powerful and we are seeing the spread of disease into new locations. These effects will only get worse culminating in an approaching environmental catastrophe.

This book has examined proposals for both *what* we should do about these challenges and *why* we should them. I have argued that existing ways of thinking about the climate – specifically as a problem in search of a "solution" – is deeply flawed. The climate cannot be stopped from changing and there is no off button, but we can influence how much (or less) it changes. So there is no end-state fixed solution, no happily ever-after. If only we consumed much less or made polluters pay for their pollution, these efforts will not prevent an environmental catastrophe in future. Such natural disasters have happened before and will happen again – whether or not there are any of us still around to see it.

But this inevitability of future catastrophe is not a reason to do nothing or see our efforts as pointless. What we do can make a very significant difference to whether any such calamity happens more quickly and its force.

We can blunt and, hopefully, delay such an event, buying ourselves precious time to adapt, mitigate and prepare.

I agree with those who say we need to ethically reframe climate change, but disagree that it need not be more than a modest redirection of the public debate.[1] Our situation leaves no room for complacency. We need to mitigate more, reducing our collective impact as much as possible. Every incremental step forward puts a future catastrophe one step back. We need to invest and develop urgently in new adaptive technologies. Their potential is enormous – and we need to ensure that benefits in areas like energy efficiency do not lead to an increasing reliance on energy.

It is popular politics in most countries to support the environment. Being green is "in." Our political leaders should not see signing a new international agreement as only an opportunity for a preelection photo-op or to seek more "likes" on social media. They should be kept to their word and drive a national – and global – commitment to preserve and maintain our environment as best we all can.

The task before us is immense. Some claim that even if could answer the big ethical questions, we may still find it too difficult to act.[2] We need to shake such doubts off. It will take political leadership for sure – but all of us must show leadership in our own ways. This sense of communal and global solidarity reinforces the bonds of collective endeavor that we require now more than ever before.

Working together

In relation to climate matters, John Donne was right to claim that our solidarity with distant strangers is possible, desirable and even necessary.[3] It is imperative that we all work together. There have been important milestones achieved in international law and policy about climate change, especially since the U.N. Framework Convention on Climate Change ratified in 1992.[4] This shows we can work across borders together on these issues. Yes, we must do more and do it more quickly. But the first steps have been taken to achieving this. The current global coronavirus COVID-19 pandemic has also shown how quick we can work together when we want to in the face of crisis. We must harness this spirit to see us rise to the challenge of the even more grave threats posed from climate change.

Much of this work operates around a Westphalian, state-centered approach. Our political leaders meet at summits to agree on new treaties and work towards new obligations. Some see the path to making progress on climate change happening only through international treaties. Climate change is global change that only a treaty signed by states will deliver.[5] This is despite widespread criticism of the relative inadequacy of our international institutions to rise to the challenge. Some, like Posner and Weisbach, claim "symbols, not substance, have been the order of the day."[6]

These same voices believe we face a choice between climate *justice* or a climate *treaty*, where the latter may not go far enough – but will go somewhere.[7] For example, Weisbach claims that "ethical arguments that ask wealthy nations to enter into treatises that systematically make them worse off violate basic feasibility constraints."[8] The focus is on feasibility putting all eggs (and hopes) in the single basket of international treatises.[9]

We can accept the importance and legal force of treaties while recognizing that not all global actors are states. In fact, some more influential international actors come from the private sector, such as multinational corporations, and other sectors including charities like the International Rescue Committee, religious organizations and civil society.[10] In working towards further achievements we should harness all such levels – and cocreate obligations on us all to work across borders and around the globe to act in our collective interests.

"No one is too small to make a difference"

We can make a difference at the sub-state level, too. International agreements might be necessary, but they are not sufficient. We make real progress in proactive efforts like the Regional Greenhouse Gas Initiative (RGGI). This is a cap-and-trade system regulating CO_2 emissions from power plants in Connecticut, Delaware, Maine, Maryland, Massachusetts, New Hampshire, New York, Rhode Island and Vermont. While it might be even better for such a system to emerge nationally, we should not let the perfect become the enemy of the good. The state is not the only body which can act.

The same goes for civil society within the state. Where states fail to discharge their duties, these sub-national actors may have a significant role to play.[11] Just as local businesses and charities do immense good supporting their local communities, these groups can – and have – play a critically important role in addressing the challenges from climate change, too. When we consider "common, but differentiated responsibilities" used as a principle at the international level, we might conceive of ways in which we might see such responsibilities taking hold at the national, sub-state and individual level. No one individual or part of the community should martyr themselves for the environmental cause; the burdens should be shared.[12] Given concerns about the inadequacy or sluggishness of national actors, our making progress on climate change is too important a task to leave to our national politicians and diplomats to undertake on their own.

Former British Prime Minister Theresa May once said that "if you believe you're a citizen of the world, you're a citizen of nowhere."[13] Her worldview is Westphalian. It is certainly true that, if possible, a strong global treaty committing all countries to action is hugely important. A next best scenario is to ensure systematic institutional change either in primary legislation at the national level or in a binding regional agreement, such as between member states of the European Union. Although these larger, structural changes can at once have a large positive impact does not mean no such impact is achievable from the bottom up. Climate change is a challenge requiring both top-down and bottom-up movements working together to secure the kinds of changes needed.

It is crucial that we all listen to diverse voices. Not only the West knows best.[14] Matters of global justice are not about applying the principles or favored frameworks from one part of the globe onto another. This is neither fair, just or respectful. We must work across cultural and political divisions as we so often do in times of emergency.

The activist Greta Thunberg argues we are living at a time of a climate emergency. She is right. Our feet stand at a crossroads. The challenges that climate change poses for us are significant. Every country, each sector and all individuals make a difference and have a role to play. The Japanese poet and writer Ryunosuke Satoro once said: "individually, we are one drop. Together, we are an ocean."[15] This message is echoed by Thunberg's claim that "No One is Too Small to Make a Difference."[16] Instead of waiting for others to act where we see the necessity of it, we must actively inspire change in others, search for new solutions and explore how we can make our contribution. Big challenges like this require a large effort.

Global Climate Fund

We have explored several proposals. There will be costs to moving to a green economy. We require some means of generating funds to cover them. There is also the issue of ensuring these efforts are properly motivated so they align with the sometimes disparate interests of different states.

We have already seen doubts about the polluter pays principle concerning how we identify polluters, how and why we make them pay and how this funding is spent. What it gets right is disincentivizing carbon emissions by increasing costs. Where it falls flat is in claiming a tax alone will be sufficient – without an overall cap – to ensure total worldwide emissions will always remain under a sustainable level. This can be easily breached as there is seemingly no limit to a polluter's pollution if he or she is willing and able to pay the price. If we hold the principle to its own standards, it does not meet its high ideals.

Nevertheless, a tax on carbon emissions has its uses for disincentivizing even greater use, as already noted, and for raising new funds that can be used for mitigation and adaptation projects alike. I want to explore a bit more how this could work better.

Let us first think about what we would want such funding to do. New monies could support carbon reduction projects, reforestation, invest in new adaptive technologies and build flood defenses. Ideally, we should support both mitigation and adaptive projects, including research and development. The aim is not about compensation, but protective preparations – shrinking our collective footprint, adapting to our changing climate and investing in our future. A small share of this funding should go to administer the fund.

One possible problem with these plans is states might want to avoid increasing their energy costs like this. Those that wish to prepare and invest can create a tax on fossil fuels within their borders. Not all might want to participate. So it is necessary to find some means of incentivizing compliance.

There are at least two ways we might do this. The first is to create a fund – let us call it the *Global Climate Fund* (GCF) – that all states pay into based on their use of fossil fuels. This can be done through a tax on the price of oil (not unlike the Global Resources Fund) or a tax linked to known emission levels discerned from a model of climate variability, if possible, and the like. (A tax on emission levels would be superior because this might incentivize the use of cleaner technologies.) This general fund provides support for projects, as already noted, concerning mitigation, adaptation and research and development. Countries paying into the GCF gain access to using the strategies and technologies supported by all contributors to the fund. This global network might provide cheaper interventions and over time build greener infrastructure within contributing states. So not only does each state get something out of paying into the fund, but they receive goods – in expertise, developing new technologies and shared practices – they cannot buy separately. This is one incentive to join the Fund.

A second incentive to join is let us say that whatever a country would be required to spend – such as $2 or so per barrel oil – half this amount would go towards projects, research and development as already noted.[17] The other half would sit with the GCF for 12 months. At year's end, each state's emission levels are assessed to see if they have decreased over that time.

If a state has not made progress on emissions reductions, then they can access this funding only through projects, research and development from the GCF – and on the same terms as they might access the first 50 percent paid into the Fund. So if a state does not make progress, it then benefits in climate change-related support for the full amount that was paid.

But if a state has made progress in reducing emissions, it gets half the money it paid into the Fund repaid. Good behavior is rewarded. And they can still benefit from the expertise and technologies developed through the common GCF.

A few caveats. First, the system is designed so that every state "wins." Those who succeed are making progress – and those who are not receive additional support. Second, the system allows states to access a collective resource delivering a collective benefit. If most countries joined this scheme, the resource available to provide the best advice, deliver the best mitigation and support the best adaptation research and development will provide benefits in excess of what any individual member pays in. Third, poorer countries use less carbon emissions per capita and so would pay in less for access than wealthier states. Fourth, the Fund would require independent governance so it is not – nor is it seen to be – on the side of any one country or region over others. Finally, it rewards consistent reductions in only paying out if progress is made. This should disincentivize inconsistent performance. A state making progress and receiving half its original payment can always invest this sum into climate change-related efforts if desired.

This is only a brief sketch of what a GCF would look like. It brings together a means by which we can ensure steady progress towards greater mitigation and less impact on the climate, it delivers a funding stream like the polluter pays principle that avoids its problems and builds in extra motivations for states to contribute and it provides a means of funding advancement in adaptive technologies. If mitigation and adaptation efforts should both be a part of our collective endeavor, the GCF is a proposal that does just that.

Light must always prevail

Climate change is an issue of great importance. Several popular policy proposals and objectives have been advocated to help us address this challenge. One problem is they do not live up to their own high ideals. These scholars – many of whom are among the brightest minds today – should not be too hard on themselves. A 2009 study examining over two decades of scientific literature on biodiversity conservation found hundreds of calls for adaptation of conservation practices "but few recommendations with sufficient specificity to inform actual operations."[18] Translating theoretical models to workable practical models is complex. Most often, these efforts have simply been overambitious – and making invaluable contributions to future policy, even if not offering us the permanent "solution" they promise. We need not choose between them, but choose how we incorporate them all as strategies for reducing our impact on the environment (and reducing the environment's impact on us).

A second problem is that the problem to be "solved" is built on a flawed foundation. A new climate change ethics is required acknowledging the continually changing climate and ever-present threat of environmental catastrophe. This is the daunting challenge we face. It is a false dichotomy to argue that a catastrophe can be avoided forever *if only* a specific policy were adopted, such as a polluter pays principle or carbon trading. None can make any such guarantee.

This does not mean that we can or should do nothing. It would be untrue to argue that unless we make substantial changes then any lesser effort is wasted. Any progress improves our position – no matter how incremental. Such a recognition does not support taking slow or minimal action instead of doing all we can. Doing something is better than nothing. However, swift and substantive action can make the difference between when an environmental catastrophe might strike – and how deadly its consequences for us and our planet. Acting now buys us time and may save lives. The inevitability of climate change and future catastrophe is a call for us to do more – not less.

In closing, we live in an endangered world. But we should not be in despair – I am a realistic optimist about the future. Achieving global climate justice is a bigger challenge than many have claimed before. We can and must provide a light in the darkness. Our activities have consequences for the planet. So, too, do our individual choices. Climate change justice requires each individual and every part of society contributing to that end. Only a positive engagement with grasping our reality and exploring the full range of possibilities can we rise to meet this greatest of all challenges. Our future waits for us to write it.

Notes

1 See Gardiner, *A Perfect Moral Storm*, 399, 404–405.
2 See Gardiner, *A Perfect Moral Storm*, 22.
3 See Henry Shue, "Distant Strangers and the Illusion of Separation: Climate, Development and Disaster," in Thom Brooks (ed.), *The Oxford Handbook of Global Justice* (Oxford: Oxford University Press, 2020): 259–276, at 272.
4 UN Framework Convention on Climate Change (UNFCCC), New York, May 9, 1992, in force March 21, 1994, 1771 UNTS 107.
5 See Gardiner, *A Perfect Moral Storm*, 95; Posner and Weisbach, *Climate Change Justice*, 2 and Stern, *A Blueprint for a Safer Planet*, 4.
6 See Posner and Weisbach, *Climate Change Justice*, 59.
7 Eric Posner, "You Can Have Either Climate Justice or a Climate Treaty, Not Both," *Slate* (November 19, 2013), url: https://slate.com/news-and-politics/2013/11/climate-justice-or-a-climate-treaty-you-cant-have-both.html.
8 Weisbach, "Introduction to Part II," 145.
9 David A. Weisbach, "Introduction to Part II," in Stephen M. Gardiner and David A. Weisbach, *Debating Climate Ethics* (Oxford: Oxford University Press, 2016): 137–169, at 142–143.
10 See Thom Brooks, "Remedial Responsibilities beyond Nations," *Journal of Global Ethics* 10 (2014): 156–166.

11 See J. Rogerlj, M. D. Elzen, N. Höhe et al., "Paris Agreement Climatic Proposals Need a Boost to Keep Warming Well Below 2°C," *Nature* 534 (2016): 631–639, at pp. 636–637 and Lachland Montgomery-Umbers and Jeremy Moss, "The Climate Duties of Sub-National Political Communities," *Political Studies* 68 (2020): 20–36.
12 See Ty Raterman, "Bearing the Weight of the World: On the Extent of an Individual's Responsibility," *Environmental Values* 21 (2012): 417–436.
13 See Theresa May, "Theresa May Conference Speech in Full," *The Daily Telegraph* (October 5, 2016), url: www.telegraph.co.uk/news/2016/10/05/theresa-mays-conference-speech-in-full/.
14 See Thom Brooks, "How Global Is Global Justice? Towards a Global Philosophy," in Thom Brooks (ed.), *New Waves in Global Justice* (Basingstoke: Palgrave Macmillan, 2014): 228–244.
15 I am grateful to Mike Bentley for sharing this quotation with me.
16 See Greta Thunberg, *No One Is Too Small to Make a Difference* (Harmondsworth: Penguin, 2019).
17 Those who think this amount of money is too low need only advocate the Global Climate Fund should raise its price.
18 Mastrandrea and Schneider, *Preparing for Climate Change*, 85.

Bibliography

Abe-Ouchi, Ayako, Fuyuki Saito, Kenki Mawamura, Maureen E. Raymo, Jun'ichi Okuno, Kunio Takashi and Heinz Blatter (2013). "Insolation-Driven 100,000-Year Glacial Cycles and Hysteresis of Ice-Sheet Volume," *Nature* 500: 190–193.

Agarwal, Anil and Sunita Narain (1991). *Global Warming in an Unequal World: A Case of Environmental Colonialism*. New Delhi: Centre for Science and Environment.

Antweiler, Werner, Brian R. Copeland and Scott Taylor (2001). "Is Free Trade Good for the Environment?" *American Economic Review* 91: 807–908.

Atkins, J. Spencer (2018). "Have You Benefitted from Carbon Emissions? You May Be a 'Morally Objectionable Free Rider'," *Environmental Ethics* 40: 283–296.

Attfield, Robin (1999). *The Ethics of the Global Environment*. Edinburgh: Edinburgh University Press.

Baer, Paul (2002). "Equity, Greenhouse Gas Emissions, and Global Common Resources," in Stephen H. Schneider, Armin Rosencranz, and John O. Niles (eds.), *Climate Change Policy: A Survey*. Washington, DC: Island Press, pp. 393–408.

Baer, Paul (2006). "Adaptation: Who Pays Whom," in Neil Adger et al. (eds.), *Fairness in Adaptation to Climate Change*. Cambridge: MIT Press, pp. 131–154.

Bailey, Ronald (ed.) (1995). *The True State of the Planet: Ten of the World's Premier Environmental Researchers in a Major Challenge to the Environmental Movement*. New York: Free Press.

Baker, Aryn (September 12, 2019). "Can a 4,815-Mile Wall of Trees Help Curb Climate Change in Africa?" *Time*, url: https://time.com/5669033/great-green-wall-africa/.

Baykal, Bilsen Beler (2019). "Recycling/Reusing Grey Water and Yellow Water: Motivations, Perspectives and Reflections into the Future," *Desalination and Water Treatment* 172: 212–223.

Bloomberg, Michael and Carl Pope (2018). *Climate of Hope: How Cities, Businesses and Citizens Can Save the Planet*. New York: St Martin's Griffin.

Booker, Christopher (2009). *The Real Global Warming Disaster*. London: Continuum.

Boyle, Alan (2012). "Human Rights and the Environment: Where Next?" *European Journal of International Law* 23: 613–642.

Boyle, Alan (2018). "Climate Change, the Paris Agreement and Human Rights," *International and Comparative Law Quarterly* 67: 759–777.

Braconnot, Pascale, Sandy P. Harrison, Masa Kageyama, Patrick J. Bartlein, Valerie Masson-Delmotte, Ayako Abe-Ouchi, Bette Otto-Bliesner and Zhao Yan (2012).

"Evaluation of Climate Models Using Palaeoclimatic Data," *Nature Climate Change* 2: 417–424.

Brand, Stewart (2010). *Whole Earth Discipline*, revised ed. London: Atlantic Books.

Brandstedt, Eric and Anna-Karin Bergman (2013)."Climate Rights: Feasible or Not?" *Environmental Politics* 22: 394–409.

Broad, William J. and Kenneth Chang (March 29, 2019). "Fossil Site Reveals Day That Meteor Hit Earth and, Maybe, Wiped Out Dinosaurs," *New York Times*, url: www.nytimes.com/2019/03/29/science/dinosaurs-extinction-asteroid.html.

Brooks, Thom (2002). "Cosmopolitanism and Distributing Responsibilities," *Critical Review of International Social and Political Philosophy* 5: 92–97.

Brooks, Thom (2007). "Punishing States That Cause Global Poverty," *William Mitchell Law Review* 33: 519–532.

Brooks, Thom (ed.) (2008). *The Global Justice Reader*. Oxford: Blackwell.

Brooks, Thom (2011). "Respect for Nature: The Capabilities Approach," *Ethics, Policy and Environment* 14: 143–146.

Brooks, Thom (2012a). "After Fukushima Daiichi: New Global Institutions for Improved Nuclear Power Policy," *Ethics, Policy and Environment* 15: 63–69.

Brooks, Thom (2012b). "Climate Change and Negative Duties," *Politics* 32: 1–9.

Brooks, Thom (ed.) (2012c). *Global Justice and International Affairs*. Boston: Brill.

Brooks, Thom (ed.) (2012d). *Justice and the Capabilities Approach*. Aldershot: Ashgate.

Brooks, Thom (2013a). "Climate Change Justice," *PS: Political Science and Politics* 46: 9–12.

Brooks, Thom (2013b). "The Real Challenge of Climate Change," *PS: Political Science and Politics* 46: 34–36.

Brooks, Thom (April, 2013c). "The Resource Curse and the Separation of Powers," *Ethics and International Affairs* (blog), url: www.ethicsandinternationalaffairs.org/2013/the-resource-curse-and-the-separation-of-powers/.

Brooks, Thom (2014a). "How Global Is Global Justice? Towards a Global Philosophy," in Thom Brooks (ed.), *New Waves in Global Justice*. Basingstoke: Palgrave Macmillan, pp. 228–244.

Brooks, Thom (2014b). "The Inevitability of Climate Change," *Global Policy* 5: 112–113.

Brooks, Thom (ed.) (2014c). *New Waves in Global Justice*. Basingstoke: Palgrave Macmillan.

Brooks, Thom (2014d). "Remedial Responsibilities beyond Nations," *Journal of Global Ethics* 10: 156–166.

Brooks, Thom (2015a). "The Capabilities Approach and Political Liberalism," in Thom Brooks and Martha C. Nussbaum (eds.), *Rawls's Political Liberalism*. New York: Columbia University Press, pp. 139–173.

Brooks, Thom (2015b). "Climate Change Justice through Taxation?" *Climatic Change* 133: 419–426.

Brooks, Thom (ed.) (2015c). *Current Controversies in Political Philosophy*. London: Routledge.

Brooks, Thom (2015d). "Why Save the Planet?" in Thom Brooks (ed.), *Current Controversies in Political Philosophy*. London: Routledge, pp. 138–147.

Brooks, Thom (2016a). "Global Justice," in Duncan Pritchard (ed.), *What Is This Thing Called Philosophy?* London: Routledge, pp. 68–80.

Brooks, Thom (2016b). "How Not to Save the Planet," *Ethics, Policy and Environment* 19: 119–135.

Brooks, Thom (2020a). "Capabilities, Freedom and Severe Poverty," in Thom Brooks (ed.), *The Oxford Handbook of Global Justice*. Oxford: Oxford University Press, pp. 199–213.

Brooks, Thom (2020b). "Climate Change Ethics and the Problem of End-State Solutions," in Thom Brooks (ed.), *The Oxford Handbook of Global Justice*. Oxford: Oxford University Press, pp. 241–258.

Brooks, Thom (ed.) (2020c). *The Oxford Handbook of Global Justice*. Oxford: Oxford University Press.

Brooks, Thom (forthcoming). "Collective Responsibility for Severe Poverty," *Global Policy*.

Brooks, Thom and Martha C. Nussbaum (eds.) (2015). *Rawls's Political Liberalism*. New York: Columbia University Press.

Broome, John (1992). *Counting the Cost of Global Warming*. Isle of Harris: White Horse Press.

Brown, Lester R. (2011). *World on the Edge: How to Prevent Environmental and Economic Collapse*. New York: W. W. Norton.

Byravan, Sujatha and Sudhir Chella Rajan (2010). "The Ethical Implications of Sea-Level Rise Due to Climate Change," *Ethics and International Affairs* 24: 239–260.

Caney, Simon (2005). "Cosmopolitan Justice, Responsibility, and Global Climate Change," *Leiden Journal of International Law* 18: 747–775.

Caney, Simon (2008). "Human Rights, Climate Change and Discounting," *Environmental Politics* 17: 536–555.

Caney, Simon and Cameron Hepburn (2011). "Carbon Trading: Unethical, Unjust and Ineffective?" *Philosophy* 69: 201–234.

Centre for Research on Energy and Clean Air (April 30, 2020). "11,000 Air Pollution-Related Deaths Avoided in Europe as Coal, Oil Consumption Plummet," url: https://energyandcleanair.org/air-pollution-deaths-avoided-in-europe-as-coal-oil-plummet/.

Corner, Adam (April 30, 2012). "Personal Carbon Allowances: A 'Big Idea That Never Took Off'," *The Guardian*, url: www.theguardian.com/sustainable-business/personal-carbon-allowances-budgets.

Crutzen, Paul J. (2006). "Albedo Enhancement by Stratospheric Sulfur Injections: A Contribution to Resolve a Policy Dilemma?" *Climatic Change* 77: 211–219.

Cyranoski, David (February 26, 2020). "Mystery Deepens over Animal Source of Coronavirus," *Nature*, url: www.nature.com/articles/d41586-020-00548-w.

Dannenberg, Astrid and Sonja Zitzelsberger (2019). "Climate Experts' Views on Geoengineering Depend on Their Beliefs about Climate Change Impacts," *Nature Climate Change* 9: 769–775.

Dobson, Andrew (2003). *Citizenship and the Environment*. Oxford: Oxford University Press.

Doran, Peter T. and Maggie Kendall Zimmerman (2009). "Examining the Scientific Consensus on Climate Change," *EOS* 90: 286–300.

Eckersley, Robyn (2010). "The Politics of Carbon Leakage and the Fairness of Border Measures," *Ethics and International Affairs* 24: 367–393.

Edelman, James (2015). "Unnecessary Causation," *Australian Law Review* 89: 20–30.

Environmental Audit Committee (2008). *Personal Carbon Trading, Fifth Report of Session 2007–08 (HC 565)*. London: House of Commons.

Environmental Protection UK (2020). "Using Wood and Coal for Home Heating," url:www.environmental-protection.org.uk/policy-areas/air-quality/air-pollution-law-and-policy/using-wood-and-coal-for-home-heating/.

Fawcett, Tina (2012). "Personal Carbon Trading: Is Now the Right Time?" *Carbon Management* 3: 283–291.

Fisher, William Arms and Katharine Lee Bates (1917). *America the Beautiful*. Boston: Oliver Ditson Company.

Fogarty, John and Michael McCally (2010). "Health and Safety Risks of Carbon Capture and Storage," *Journal of the American Medical Association* 303: 67–68.

Gaines, S. (1991). "The Polluter-Pays Principle: From Economic Equity to Environmental Ethos," *Texas International Law Journal* 26: 463–495.

Garcia, Jorge H. and Asbjørn Torvanger (2019). "Carbon Leakage from Geological Storage Sites: Implications for Carbon Trading," *Energy Policy* 127: 320–329.

Gardiner, Stephen M. (2004). "Ethics and Global Climate Change," *Ethics* 114: 555–600.

Gardiner, Stephen M. (2011). *A Perfect Moral Storm: The Ethical Tragedy of Climate Change*. Oxford: Oxford University Press.

Gardiner, Stephen M. (2016). "Betraying the Future," in Stephen M. Gardiner and David A. Weisbach (eds.), *Debating Climate Ethics*. Oxford: Oxford University Press, pp. 6–45.

Garvey, James (2008). *The Ethics of Climate Change: Right and Wrong in a Warming World*. London: Continuum.

Giddens, Anthony (2009). *The Politics of Climate Change*. Cambridge: Polity.

Gore, Albert (1992). *Earth in the Balance: Ecology and the Human Spirit*. Boston: Houghton Mifflin.

Grantham Research Institute on Climate Change and the Environment (May 17, 2018). "What Is a Carbon Price and Why Do We Need One?" url: www.lse.ac.uk/GranthamInstitute/faqs/what-is-a-carbon-price-and-why-do-we-need-one/.

Greiner, D. James (2008). "Causal Inference in Civil Rights Litigation," *Harvard Law Review* 122: 533–598.

Haller, Stephen F. (2002). *Apocalypse Soon? Wagering on Warnings of Global Catastrophe*. Montreal: McGill-Queens.

Hardin, Garrett (1968). "The Tragedy of the Commons," *Science* 162: 1243–1248.

Harris, Paul G. (2010). *World Ethics and Climate Change: From International to Global Justice*. Edinburgh: Edinburgh University Press.

Hasselmann, Klaus (1997). "Multi-Pattern Fingerprint Method for Detection and Attribution of Climate Change," *Climate Dynamics* 13: 601–611.

Hassoun, Nicole (2011). "The Anthropocentric Advantage? Environmental Ethics and Climate Change Policy," *Critical Review of International Social and Political Philosophy* 14: 235–257.

Hauer, Mathew E., Jason M. Evans and Deepak R. Mishra (2016). "Millions Projected to Be at Risk from Sea-Level Rise in the Continental United States," *Nature Climate Change* 6: 691–695.

Hayward, Tim (2005). *Constitutional Environmental Rights*. Oxford: Oxford University Press.

Hazeldine, R. Stuart (2009). "Carbon Capture and Storage: How Green Can Black Be?" *Science* 325: 1647–1652.

Hegerl, G. C., Klaus Hasselmann, U. Cubasch, J. F. B. Mitchell, E. Roeckner, R. Voss and J. Waszkewitz (1997). "Multi-Fingerprint Detection and Attribution Analysis of Greenhouse Gas, Greenhouse Gas-Plus-Aerosol and Solar Forced Climate Change," *Climate Dynamics* 13: 613–634.

Henberg, M. C. (1977). "Nozick and Rawls on Historical versus End-State Distribution," *Southwestern Journal of Philosophy* 8: 77–84.

Hepburn, Cameron (2007). "Carbon Trading: A Review of the Kyoto Mechanisms," *Annual Review of Environmental Resources* 32: 375–393.

Hepburn, Cameron and Nicholas Stern (2009). "The Global Deal on Climate Change," in Dieter Helm and Cameron Hepburn (eds.), *The Economics and Politics of Climate Change*. Oxford: Oxford University Press, pp. 36–57.

Hillman, Mayer (2008). *How We Can Save the Planet: Preventing Global Climatic Catastrophe*. New York: St Martin's Press.

Hoekstra, Arjen Y. and Mesfin M. Mekonnen (201). "The Water Footprint of Humanity," *Proceedings of the National Academy of Science* 109: 3232–3237.

Houghton, John (1997). *Global Warming: The Complete Briefing*, 2nd ed. Cambridge: Cambridge University Press.

Hugo, Victor (1907). *The Works of Victor Hugo: Les Miserables*, vol. 7. London: Jensen Society.

Hulme, Mike (2009). *Why We Disagree about Climate Change: Understanding Controversy, Inaction and Opportunity*. Cambridge: Cambridge University Press.

Intergovernmental Panel on Climate Change (2014). *Climate Change 2014: Impacts, Adaptation, and Vulnerability*. Cambridge: Cambridge University Press.

Intergovernmental Panel on Climate Change (2019). "Special Report on the Ocean and Cryosphere in a Changing Climate: Summary for Policymakers," url: www.ipcc.ch/site/assets/uploads/sites/3/2019/11/03_SROCC_SPM_FINAL.pdf.

Intergovernmental Panel on Climate Change (2020). *Special Report: Global Warming of 1.5°C*. Geneva: Intergovernmental Panel on Climate Change.

International Committee of the Red Cross (September 26, 2019). "A Drought So Severe It Has a Name," *International Committee of the Red Cross*, url: www.icrc.org/en/document/somalia-conflict-drought-so-severe-it-has-names.

International Organization for Migration (2019). *World Migration Report 2020*. Geneva: International Organization for Migration.

Jacobson, Mark Z. (2019). "The Health and Climate Impacts of Carbon Capture and Direct Air Capture," *Energy and Environmental Science* 12: 3567–3574.

Jameton, Andrew (2016). "Time Frames for Saving the Planet," *Ethics, Policy and Environment* 19: 136–140.

Jamieson, Dale (1991). "The Epistemology of Climate Change: Some Morals for Managers," *Society and Natural Resources* 4: 319–329.

Jamieson, Dale (2008). *Ethics and the Environment*. Cambridge: Cambridge University Press.

Jamieson, Dale (2014). *Reason in a Dark Time: Why the Struggle against Climate Change Failed: And What It Means for Our Future*. Oxford: Oxford University Press.

Ju, Biung-Ghi and Juan D. Moreno-Ternero (2018). "Entitlement Theory of Justice and End-State Fairness in the Allocation of Goods," *Economics and Philosophy* 34: 317–341.

Kahn, Matthew E. (2010). *Climatopolis: How Our Cities Will Thrive in the Hotter Future*. New York: Basic Books.

King, David A. (2004). "Climate Change Science: Adapt, Mitigate, or Ignore?" *Science* 9: 176–177.

Knapp, Christopher (2011). "Tragedies without Commons," *Public Affairs Quarterly* 25: 81–94.

Kolers, Avery (2012). "Floating Provisos and Sinking Islands," *Journal of Applied Philosophy* 29: 333–343.

Kurtzman, J. (2009). "The Low Carbon Diet," *Foreign Policy* 88: 114–122.

Kyoto Protocol to the UN Framework on Climate Change, Kyoto, Japan, December 10, 1997, in force February 16, 2005, 2303 UNTS 148.

Lazarowicz, M. (2009). *Global Carbon Trading: A Framework for Reducing Emissions*. London: TSO.

Lee, Alexander and Jordan Kincaid (2016). "Two Problems of Climate Ethics: Can We Lose the Planet but Save Ourselves?" *Ethics, Policy and Environment* 19: 141–144.

Leijten, Ingrid (2019). "Human Rights versus Insufficient Climate Change: The Urgenda Case," *Netherlands Quarterly of Human Rights* 37: 112–118.

Levitt, Steven D. and Stephen J. Dubner (2010). *Superfreakonomics*. London: Penguin.

Li, Yao, Jin Fan, Dingtao Zhao, Yanrui Wu and Jun Li (2016). "Tiered Gasoline Pricing: A Personal Carbon Trading Perspective," *Energy Policy* 89: 194–201.

Liao, S. Matthew, Anders Sandberg and Rebecca Roache (2012). "Human Engineering and Climate Change," *Ethics, Policy and Environment* 15: 206–221.

Lomborg, Bjorn (1998). *The Skeptical Environmentalist: Measuring the Real State of the World*. Cambridge: Cambridge University Press.

Lomborg, Bjorn (2008). *Cool It: The Skeptical Environmentalist's Guide to Global Warming*. New York: Vintage.

Loo, Clement (2016). "Environmental Justice as a Foundation for a Process-Based Framework for Adaptation and Mitigation: A Commentary on Brooks," *Ethics, Policy Environment* 19: 145–149.

Lopez, Juan Miguel Rodriguez, Anita Engels and Lisa Knoll (2017). "Understanding Carbon Trading: Effects of Delegating CO_2 Responsibility on Organizations' Trading Behaviour," *Climate Policy* 17: 346–360.

Lovelock, James (2000). *The Ages of Gaia: A Biography of Our Living Earth*, 2nd ed. Oxford: Oxford University Press.

Lute, Michelle L., Shahzeen Z. Attari and Steven J. Sherman (2015). "Don't Rush to Flush," *Journal of Environmental Psychology* 43: 105–111.

MacDonald, Kerri (October 18, 2011). "As Water Rises, There's No Place Like (or for) Home," *New York Times*, url: https://lens.blogs.nytimes.com/2011/10/18/no-place/.

Malnes, Raino (2008). "Climate Science and the Way We Ought to Think about Danger," *Environmental Politics* 17: 660–672.

Marshall, Shawn J. (2013). "Climate Science: Solution Proposed for Ice-Age Mystery," *Nature* 500: 159–160.

Mastrandrea, Michael D. and Stephen H. Schneider (2010). *Preparing for Climate Change*. Cambridge: MIT Press.

May, Theresa (October 5, 2016). "Theresa May Conference Speech in Full," *The Daily Telegraph*, url: www.telegraph.co.uk/news/2016/10/05/theresa-mays-conference-speech-in-full/.

Mazar, Joseph (2010). "Liberal Justice, Future People, and Natural Resource Conservation," *Philosophy and Public Affairs* 38: 380–408.

McLean, Elena V. and Tatyana Plaksina (2019). "The Political Economy of Carbon Capture and Storage Technology Adoption," *Global Environmental Politics* 19: 127–148.

Mill, John Stuart (1978). *On Liberty*. Indianapolis: Hackett.

Minnerop, Petra (2019). "Integrating the 'Duty of Care' under the European Convention on Human Rights and the Science and Law of Climate Change: The Decision of the Hague Court of Appeal in the Urgenda Case," *Journal of Energy and Natural Resources Law* 37: 149–179.

Moellendorf, Darrel (2009). "Treaty Norms and Climate Change Mitigation," *Ethics and International Affairs* 23: 247–265.

Montgomery-Umbers, Lachland and Jeremy Moss (2020). "The Climate Duties of Sub-National Political Communities," *Political Studies* 68: 20–36.

Mooney, Chris (January 4, 2018). "How Climate Change Could Counterintuitively Feed Winter Storms," *Washington Post*, url: www.washingtonpost.com/news/energy-environment/wp/2018/01/04/how-climate-change-could-counterintuitively-feed-some-winter-storms/.

Mylius, Ben (2016). "Change-Oriented Conceptions of Climate: A Response to Thom Brooks' How Not to Save the Planet," *Ethics, Policy and Environment* 19: 150–152.

NASA (2020a). "The Causes of Climate Change," url: https://climate.nasa.gov/causes/.

NASA (2020b). "Overview: Weather, Global Warming and Climate Change," url: https://climate.nasa.gov/resources/global-warming-vs-climate-change/.

Neumayer, Eric (2000). "In Defence of Historical Accountability for Greenhouse Gas Emissions," *Ecological Economics* 33: 185–192.

Nine, Cara (2010). "Ecological Refugees, States Borders, and the Lockean Proviso," *Journal of Applied Philosophy* 27: 359–375.

Nozick, Robert (1974). *Anarchy, State and Utopia*. New York: Basic Books.

Nunn, Patrick D. (September 11, 2017). "Sinking Islands: Sea Level Rise Is Washing away Micronesia's History," *Newsweek*, url: www.newsweek.com/sea-level-rise-vanishing-islands-micronesia-history-706455.

Nussbaum, Martha C. (2000). *Women and Human Development: The Capabilities Approach*. Cambridge: Cambridge University Press.

O'Neill, Eoin (2016). "The Precautionary Principle: A Preferred Approach for the Unknown," *Ethics, Policy and Environment* 19: 153–156.

O'Neill, Onora (2001). "Agents of Justice," *Metaphilosophy* 32: 180–195.

Pacala, S. and R. Socolow (2004). "Stabilization Wedges: Solving the Climate Problem for the Next 50 Years with Current Technologies," *Science* 305: 968–972.

Page, Edward A. (2006). *Climate Change, Justice, and Future Generations*. Cheltenham: Edward Elgar.

Page, Edward A. (2011a). "Cashing in on Climate Change: Political Theory and Global Emissions Trading," *Critical Review of International Social and Political Philosophy* 14: 1–15.

Page, Edward A. (2011b). "Cosmopolitanism, Climate Change, and Greenhouse Gas Emissions Trading," *International Theory* 3: 37–69.

Parfit, Derek (1984). *Reasons and Persons*. Oxford: Oxford University Press.

Park, Jongkwan et al. (2019). "Organic Matter Composition of Manure and Its Potential Impact on Plant Growth," *Sustainability* 11 (2346): 1–12.

Parry, Martin L., Nigel W. Arnell, Anthony J. McMichael, Robert J. Nicholls, Pim Martens, R. Sari Kovats, Matthew T. J. Livermoore, Cynthia Rosenzweig, Ana Iglesias and Gunther Fischer (2001). "Millions at Risk: Defining Critical Climate Change Threats and Targets," *Global Environmental Change: Human and Policy Dimensions*: 181–183.

Pogge, Thomas (2008). *World Poverty and Human Rights*, 2nd ed. Cambridge: Polity.

Pollin, Robert (2020). "The Green Growth Path to Stabilization," in Akeel Bilgrami (ed.), *Nature and Value*. New York: Columbia University Press, pp. 117–126.

Posner, Eric (November 19, 2013). "You Can Have Either Climate Justice or a Climate Treaty, Not Both," *Slate*, url: https://slate.com/news-and-politics/2013/11/climate-justice-or-a-climate-treaty-you-cant-have-both.html.

Posner, Eric and Cass Sunstein (2009). "Should Greenhouse Gas Permits Be Allocated on a Per Capita Basis?" *California Law Review* 97: 51–93.

Posner, Eric A. and David Weisbach (2010). *Climate Change Justice*. Princeton: Princeton University Press.

Posner, Richard A. (2004). *Catastrophe: Risk and Response*. Oxford: Oxford University Press.

Rajamani, Lavanja (2010). "The Increasing Currency and Relevance of Rights-Based Perspectives in the International Negotiations on Climate Change," *Journal of Environmental Law* 22: 391–429.

Ratcliffe, Rebecca (June 6, 2019). "Two Million People at Risk of Starvation as Drought Returns to Somalia," *The Guardian*, url:www.theguardian.com/global-development/2019/jun/06/two-million-people-at-risk-of-starvation-as-drought-returns-to-somalia.

Raterman, Ty (2012). "Bearing the Weight of the World: On the Extent of an Individual's Responsibility," *Environmental Values* 21: 417–436.

Raup, David M. and J. John Sepkoski, Jr. (1982). "Mass Extinctions in the Marine Fossil Record," *Science* 215: 1501–1503.

Rees, William E. (1992). "Ecological Footprints and Appropriated Carrying Capacity: What Urban Economics Leaves Out," *Environment and Urbanization* 4: 121–130.

Rose, C. M. (2000). "Expanding the Choices for the Global Commons: Comparing Newfangled Tradable Allowance Schemes to Old-Fashioned Common Property Regimes," *Duke Environmental Law and Policy Forum* 10: 45–72.

Roy, Eleanor Ainge (May 16, 2019). "'One Day We'll Disappear': Tuvalu's Sinking Islands," *The Guardian*, url: www.theguardian.com/global-development/2019/may/16/one-day-disappear-tuvalu-sinking-islands-rising-seas-climate-change.

Royal Society (March, 2020). "How Fast Is Sea Level Rising," url: https://royalsociety.org/topics-policy/projects/climate-change-evidence-causes/question-14/?gclid=EAIaIQobChMIqsC5hc7m6AIVE-3tCh3r2A4-EAAYASAAEgJQR_D_BwE.

Risse, Mathias (2009). "The Right to Relocation: Disappearing Island Nations and Common Ownership of the Earth," *Ethics and International Affairs* 23: 281–300.

Rogerlj, J., M. D. Elzen, N. Höhe et al. (2016). "Paris Agreement Climatic Proposals Need a Boost to Keep Warming Well Below 2°C," *Nature* 534: 631–639.

Roser, Dominic and Luke Tomlinson (2014). "Trade Policies and Climate Change: Border Carbon Adjustments as a Tool for a Just Global Climate Regime," *Ancilla Iuris*: 222–244.

Sagoff, Mark (1988). *The Economy of the Earth*. Cambridge: Cambridge University Press.

Sagoff, Mark (2002). "Controlling Global Climate: The Debate over Pollution Trading," in V. V. Gehring and W. A. Galston (eds.), *Philosophical Dimensions of Public Policy*. London: Transaction Publishers, pp. 311–318.

Sandel, Michael (2005). *Public Philosophy: Essays on Morality in Politics*. Cambridge: Harvard University Press.

Schmidtz, David (2001). "A Place for Cost-Benefit Analysis," *Noûs* 11 (supplement): 148–171.

Schumacher, E. F. (1973). *Small Is Beautiful*. New York: Harper & Row.

Shell, Jonathan (2020). "Nature and Value," in Akeel Bilgrami (ed.), *Nature and Value*. New York: Columbia University Press, pp. 1–12.

Shimkus, David (October 31, 2017). "Addressing Climate Change Means Address ing the Global Sanitation Crisis," *International Institute for Sustainable Develop ment*, url: https://sdg.iisd.org/commentary/guest-articles/addressing-climate-change-means-addressing-the-global-sanitation-crisis/.

Shorrocks, Antony, Jim Davies and Rodrigo Lluberas (2019). *Global Wealth Report 2019*. Geneva: Credit Suisse.

Shue, Henry (1993). "Subsistence Emissions and Luxury Emissions," *Law and Policy* 15: 39–59.

Shue, Henry (1999). "Global Environment and International Inequality," *International Affairs* 75: 533–537.

Shue, Henry (2020). "Distant Strangers and the Illusion of Separation: Climate, Development and Disaster," in Thom Brooks (ed.), *The Oxford Handbook of Global Justice*. Oxford: Oxford University Press, pp. 259–276.

Singer, Peter (2004). *One World: The Ethics of Globalization*, 2nd ed. New Haven, CT: Yale University Press.

Sky News (April 20, 2020). "Coronavirus: US Oil Price Plunges below Zero for First Time in History as Outbreak Hits Demand," url: https://news.sky.com/story/us-oil-price-plunges-below-zero-for-first-time-in-history-11976165.

Stapleton, Jane (2015). "An 'Extended-but-for' Test for the Causal Relation in the Law of Obligations," *Oxford Journal of Legal Studies* 35: 697–726.

Stavins, R. N. (2008). "Addressing Climate Change with a Comprehensive US Cap-and-Trade System," *Oxford Review of Economic Policy* 24: 298–321.

Stern, Nicholas (2009). *A Blueprint for a Safer Planet*. London: Bodley Head.

Sunstein, Cass R. (2005). *Laws of Fear: Beyond the Precautionary Principle*. Cambridge: Cambridge University Press.

Svoboda, Toby, Holly Buck and Pablo Suarez (2019). "Climate Engineering and Human Rights," *Environmental Politics* 28: 397–416.

Thunberg, Greta (2019). *No One Is Too Small to Make a Difference*. Harmondsworth: Penguin.

Tietenberg, T. (2006). *Emissions Trading: Principles and Practice*, 2nd ed. Washington, DC: Resources for the Future.

Tilton, John E. (2016). "Global Climate Policy and the Polluter Pays Principle: A Different Perspective," *Resources Policy* 50: 117–118.

Traxler, Martino (2002). "Fair Chore Division for Climate Change," *Social Theory and Practice* 28: 101–134.

Trump, Donald J. (December 29, 2017). "Tweet," url: https://twitter.com/realDonaldTrump/status/946531657229701120.

UN Environment Programme (2020a). "Climate Adaptation," url: www.unep.org/explore-topics/climate-change/what-we-do/climate-adaptation.

UN Environment Programme (2020b). "Facts about Climate Emergency," url: www.unenvironment.org/explore-topics/climate-change/facts-about-climate-emergency.

UN Environment Programme (2020c). "Mitigation," url: www.unep.org/explore-topics/climate-change/what-we-do/mitigation.

U.S. Energy Information Administration (2011). *International Energy Outlook 2011*. Washington, DC: U.S. Energy Information Administration, pp. 6–7, url: www.eia.gov/forecasts/ieo/emissions.cfm/.

Van den Bergh, Jeroen C. J. M. and Harmen Verbruggen (1999). "Spatial Sustainability, Trade, Trade, and Indicators: An Evaluation of the 'Ecological Footprint'," *Ecological Economics* 29: 63–74.

Vanderheiden, Steve (2008). "Two Conceptions of Sustainability," *Political Studies* 56: 435–455.

Wackernagel, Mathis (2009). "Methodological Advancements in Footprint Analysis," *Ecological Economics*: 1925–1927.

Wackernagel, Mathis and William E. Rees (1996). *Our Ecological Footprint: Reducing Human Impact on the Earth*. Gabriola Island: New Society Publishers.

Wackernagel, Mathis and William E. Rees (1997). "Perceptual and Structural Barriers to Investing in Natural Capital: Economics from an Ecological Footprint Perspective," *Ecological Economics* 20: 3–24.

Walsh, Nick Paton and Vasco Cotovio (March 20, 2020). "Bats Are Not to Blame for Coronavirus: Humans Are," *CNN*, url: https://edition.cnn.com/2020/03/19/health/coronavirus-human-actions-intl/index.html?iid=ob_article_organicsidebar_expansion.

Warne, Kennedy (November–December, 2004). "That Sinking Feeling," *New Zealand Geographic* 70, url: www.nzgeo.com/stories/tuvalu/.

Wasike, Andrew (September 18, 2019). "World in Progress: Somalia's Climate Refugees," *Deutsche Welle*, url: www.dw.com/en/world-in-progress-somalias-climate-refugees/av-50480611.

Watts, Jonathan (April 9, 2020). "Climate Crisis: In Coronavirus Lockdown, Nature Bounces Back: But for How Long?" *The Guardian*, url: www.theguardian.com/

world/2020/apr/09/climate-crisis-amid-coronavirus-lockdown-nature-bounces-back-but-for-how-long.

Weber, Rolf H. (2015). "Border Tax Adjustments: Legal Perspective," *Climatic Change* 133: 407–417.

Weisbach, David A. (2016a). "Climate Policy and Self-Interest," in Stephen M. Gardiner and David A. Weisbach (eds.), *Debating Climate Ethics*. Oxford: Oxford University Press, pp. 170–200.

Weisbach, David A. (2016b). "Introduction to Part II," in Stephen M. Gardiner and David A. Weisbach (eds.), *Debating Climate Ethics*. Oxford: Oxford University Press, pp. 137–169.

Wenar, Leif (2016). *Blood Oil: Tyrants, Violence and the Rules That Run the World*. Oxford: Oxford University Press.

Woollings, Tim (November 21, 2019). "How Climate Change Could Be Affecting the Jet Stream," *The Independent*, url: www.independent.co.uk/news/science/climate-change-crisis-latest-jet-stream-extreme-weather-a9208901.html.

World Health Organization (2020). "Coronavirus Disease (COVID-19) Pandemic," url: www.who.int/emergencies/diseases/novel-coronavirus-2019.

World Wildlife Fund (2020). "How Big Is Your Environmental Footprint?" url: https://footprint.wwf.org.uk/.

Yuichi Kono, Daniel (2020). "Compensating for the Climate: Unemployment Insurance and Climate Change Votes," *Political Studies* 68: 167–186.

Index